Springer Series in Reliability Engineering

Series Editor

Professor Hoang Pham
Department of Industrial and Systems Engineering
Rutgers, The State University of New Jersey
96 Frelinghuysen Road
Piscataway, NJ 08854-8018
USA

Other titles in this series

The Universal Generating Function in Reliability Analysis and Optimization
Gregory Levitin

Warranty Management and Product Manufacture
D.N.P. Murthy and Wallace R. Blischke

Maintenance Theory of Reliability
Toshio Nakagawa

System Software Reliability
Hoang Pham

Reliability and Optimal Maintenance
Hongzhou Wang and Hoang Pham

Applied Reliability and Quality
B.S. Dhillon

Shock and Damage Models in Reliability Theory
Toshio Nakagawa

Risk Management
Terje Aven and Jan Erik Vinnem

Satisfying Safety Goals by Probabilistic Risk Assessment
Hiromitsu Kumamoto

Offshore Risk Assessment (2nd Edition)
Jan Erik Vinnem

The Maintenance Management Framework
Adolfo Crespo Márquez

Human Reliability and Error in Transportation Systems
B.S. Dhillon

Complex System Maintenance Handbook
D.N.P. Murthy and Khairy A.H. Kobbacy

Recent Advances in Reliability and Quality in Design
Hoang Pham

Product Reliability
D.N.P. Murthy, Marvin Rausand and Trond Østerås

Mining Equipment Reliability, Maintainability, and Safety
B.S. Dhillon

Advanced Reliability Models and Maintenance Policies
Toshio Nakagawa

Justifying the Dependability of Computer-based Systems
Pierre-Jacques Courtois

Reliability and Risk Issues in Large Scale Safety-critical Digital Control Systems
Poong Hyun Seong

Failure Rate Modeling for Reliability and Risk
Maxim Finkelstein

The Complexity of Proceduralized Tasks
Jinkyun Park

Risks in Technological Systems
Göran Grimvall, Åke J. Holmgren, Per Jacobsson and Torbjörn Thedéen

Maintenance for Industrial Systems
Riccardo Manzini, Alberto Regattieri, Hoang Pham and Emilio Ferrari

B.S. Dhillon

Mine Safety

A Modern Approach

Springer

B.S. Dhillon, PhD
Department of Mechanical Engineering
University of Ottawa
Ottawa
Ontario K1N 6N5
Canada

ISSN 1614-7839
ISBN 978-1-84996-114-1 e-ISBN 978-1-84996-115-8
DOI 10.1007/978-1-84996-115-8

British Library Cataloguing in Publication Data
A catalogue record for this book is available from the British Library

Library of Congress Control Number: 2010925893

Cover design: deblik, Berlin, Germany

Printed on acid-free paper

Springer is part of Springer Science+Business Media (www.springer.com)

This book is affectionately dedicated to
Mr. Harry Sumner of England
for his inspirational comments,
faith in the author's capability,
and help in various areas.

Preface

The mining industry has played a pivotal role in the development of civilization. The history of mining may be traced back to ancient Egyptians who operated malachite mines. Today the world mining industry is producing over 6 billion tons of raw product annually that is worth trillions of dollars.

In recent years, mine safety has become an important issue because each year thousands of miners die from mining accidents around the world. These accidents have a variety of causes including dust explosions, collapsing of mine stopes, flooding, and general mechanical errors from improperly used or malfunctioning mining equipment or systems.

Although a large number of journal and conference proceedings articles on mine safety have appeared, to the best of author's knowledge there is no book on the topic that covers recent developments in the area. This causes a great deal of difficulty for those who seek information on the subject because they have to consult many different and diverse sources.

The main objective of this book is to consolidate the desired safety information and to provide up-to-date information on the subject. The sources of most of the material presented are given in the reference sections at the end of each chapter. This will be useful to readers if they desire to delve deeper into a specific area. The book contains a chapter on safety mathematics and basics, and two chapters on safety management and safety analysis methods and indices considered essential to the understanding of the contents of subsequent chapters.

The topics covered in the volume are treated in such a manner that the reader will require no previous knowledge to understand the contents. At appropriate places, the book contains examples along with their solutions, and at the end of each chapter there are numerous problems to test reader comprehension.

An extensive list of references on mine safety is provided at the end of the book to give readers a view of developments on the subject over the years.

The book is composed of 11 Chapters. Chapter 1 presents the historical developments in mine safety, the need for improving safety in mining, mine safety facts and figures, major mine disasters, important terms and definitions, the overall scope of the book, and useful information on mine safety classified under five distinct categories. Chapter 2 is devoted to safety mathematics and basics, and covers topics such as standard deviation, Boolean algebra laws and probability properties, probability distributions, Laplace transform and expected value definitions, statute, administrative, common, and liability laws, accident causation theories, and common causes of work injuries.

Chapter 3 presents various important aspects of safety management including safety management principles and safety department functions, safety manager's and engineer's functions, safety-related strategies for safety professionals, management-related deficiencies leading to accidents, and safety checklists for management. Chapter 4 is devoted to safety analysis methods and indices and covers items such as hazards and operability analysis (HAZOP), job safety analysis (JSA), interface safety analysis (ISA), failure mode and effect analysis (FMEA), fault tree analysis (FTA), and disabling injury severity rate (DISR).

A number of global mine accidents are covered in Chapter 5. Some of these accidents are the Monongah mining disaster, the Cherry mine disaster, the Senghenydd colliery disaster, the Sunjiawan mine disaster, the Bulli colliery disaster, the Hillcrest mining disaster, and the Ulyanovskya coal mine disaster.

Chapters 6 and 7 are devoted to human factors and error in mine safety and mining equipment safety, respectively. Chapter 6 covers topics such as the need for the application of human factors in the mining industry, occupational stressors, human factor-safety issues, classifications and causes of human errors resulting in fatal mine accidents, common mining equipment maintenance errors, maintenance error contributory factors, methods for performing human error analysis in the area of mine safety, and factors responsible for failing to reduce the occurrence of human error in the mining sector. Some of the topics covered in Chapter 7 are equipment fire-related mining accidents, mining equipment fire ignition sources, strategies for reducing mining equipment fires, safety in electrical design of mine elevator control systems, guidelines for improving electrical safety in mines, mining equipment safety analysis methods, and hazardous area signalling and ranging device (HASARD) proximity warning systems.

Chapter 8 presents various important aspects of electrical accidents in mines and programmable electronic mining system safety. These include fatal electrical accidents in comparison to other fatal mine accidents and in different mining areas, job titles of victims of mining electrical accidents, measures for mitigating mine electrical shock injuries in general and in maintenance work, programmable electronic-related accidents in mines, methods for conducting programmable electronic mining system hazard and risk analysis, and lessons learned in addressing programmable electronic mining system safety-related issues.

Chapter 9 is devoted to gas-related, fire, and blasting accidents in mines and methods for determining mine atmosphere status. It covers topics such as origins and mechanisms of coal mine outbursts and their prediction and prevention, underground fires in hard coal mines, use of inert gases in mine fires, blasting injuries in surface mining and blast damage index, and methods for determining the status of mine atmosphere. Various important aspects of safety in offshore industry are covered in Chapter 10. Some of the topics covered include offshore risk picture, offshore accident causes, accident reporting approach in the offshore industry, and case studies of accidents in offshore industry. Finally, Chapter 11 presents eight mathematical models considered useful for performing safety analysis in the area of mining.

The book will be useful to many individuals including engineering and safety professionals working in the mining industry, mining administrators, mining engineering undergraduate and graduate students, researchers and instructors in the area of mining, and human factor and safety specialists-at-large.

The author is deeply indebted to many individuals, including students, colleagues, and friends for their invisible inputs. The invisible contributions of my children, Jasmine and Mark, are also appreciated. Last, but not least, I thank my boss, friend, and wife, Rosy for typing this book and for her timely help in proofreading.

Ottawa, Ontario, March 2010 B.S. Dhillon

Contents

Chapter 1
Introduction

1.1 Historical Developments in Mine Safety

Although the history of mining may be traced back to the ancient Egyptians who operated malachite mines at Wady Maghareh on the Sinai Peninsula and at Timna in the Negev, it was not until 1891 when the United States Congress passed the first federal statute governing the area of mine safety [1, 2]. In 1910, the United States Congress established the Bureau of Mines as a new agency to look after mine-related matters within the framework of the Department of the Interior.

This bureau was charged with the responsibility to perform research and to reduce the occurrence of accidents in the coal mining industrial sector, but it was not provided with any inspection authority until 1941, when Congress empowered federal inspection personnel to enter mines [1]. In 1947, Congress authorized the formulation of the first code of federal regulations in the area of mine safety and the Federal Coal Mine Safety Act of 1952 provided for yearly inspections in specific underground coal mines, and gave the Bureau of Mines a certain degree of enforcement authority.

In 1966, the 1952 Coal Mine Safety Act was extended to all underground coal mines by the Congress. In 1969, the Congress passed the Coal Mine Safety and Health Act, which was more stringent and comprehensive than any earlier Federal Legislation governing the mining industrial sector in the United States [1]. In 1973, a new departmental agency separate from the Bureau of Mines called the Mining Enforcement and Safety Administration (MESA) was created by the Department of the Interior. Prior to March, 1978 MESA was the predecessor organization to Mine Safety and Health Administration (MSHA) [1, 3].

In 1977, the Federal Mine Safety and Health Act was passed by Congress, which is the legislation that governs all MSHA's activities. This Act has helped to reduce mining fatalities quite considerably over the years in the United States (i.e., 272 in 1977 to less than 100 in 2007) [1].

Over the years many publications covering various types of developments in the area of mine safety have appeared. A comprehensive list of these publications is provided at the end of this book.

1.2 Need for Approving Safety in Mining

Each year world mining industries produce billions of tons of raw product worth trillions of dollars and employs millions of people worldwide. For example, the United States natural resources and mining sector alone employs around 675,000 people [4]. Every year thousands of people are killed and injured in the mining sector throughout the world.

Mining fatalities and injuries are declining in certain parts of the world, but they are still challenging issues. For example, in the United States mining fatalities decreased from over 3,000 per year in 1910 to 55 in 2004 [5]. However, the risk of being killed in mining in America is still roughly six times greater than that of the general industrial sector and the risk of being injured is double that of other industrial sectors. All in all, it simply means that there is still a definite need to improve safety in the area of mining.

1.3 Mine Safety Facts and Figures

Some facts and figures that are directly or indirectly concerned with mine safety are as follows:

- In 2000, there were around 5200 fatalities due to work-related accidents in the United States [6] and in 1995 work accidents cost the United States roughly $75 billion [7].
- In 2006, 72 miners were killed in the United States, 47 of which were in the coal mining sector [8, 9].
- During the period 1990–1999, electricity was the fourth leading cause of fatalities in the United States mining industrial sector [10].
- During the period 1983–1992, 81 deaths occurred in quarries throughout the United Kingdom [11].
- Annual mining fatalities in the United States decreased from about 500 in the late 1950s to around 93 during the 1990s [12].
- During the period 1978–1988, maintenance activity accounted for around 34% of all lost-time injuries in the United States mining industrial sector [13].
- Over 25% of the accidents in underground coal mines occurred during maintenance activity [14].
- During the period 1990–1999, a total of 197 equipment fires caused 76 injuries in coal mining operations throughout the United States [15].

- In 2004, about 17% of the 37,445 injuries in the United States coal mines were directly or indirectly associated with bolting machines [16].
- During the period 1991–1999, an average of 21,351 mining-associated injuries per year occurred in the United States [12].
- During the period 1992–2001, the average fatality rate in the United States mining industrial sector was 27.7 per 100,000 workers, as opposed to 4.8 per 100,000 workers in all United States industries [17].
- During the period 1983–1990, as per injury data for independent contractor employees in the United States mining industrial sector, about 20% of the coal mine-associated injuries occurred during equipment maintenance or while using hand tools [18].

1.4 Major Mine Disasters

Over the years many mine disasters have occurred around the globe. Some of the major ones are as follows:

- **Benxihu (Honkeiko) Colliery Mining Disaster.** This disaster occurred at the colliery located near Benxi Lake in the Ore-rich region of eastern Chinese province called Lianoning on April 26, 1942 and killed 1549 miners [19].
- **Senghenydd Colliery Disaster.** This disaster occurred at the Senghenydd Colliery in Senghenydd, South Wales, United Kingdom on October 14, 1913 and killed 439 people [20, 21].
- **Coalbrook Coal Mine Disaster.** This disaster occurred at the Coalbrook Coal Mine in South Africa in 1960 and caused 437 fatalities [22].
- **The Oaks Mining Disaster.** This disaster occurred at the Oaks Mine in Yorkshire, England, United Kingdom on December 12, 1866 and killed 388 miners [23].
- **Monongah Mining Disaster.** This disaster occurred in Monongah, West Virginia on December 6, 1907 and has been described as "the worst mining disaster in the history of the United States", because it caused 362 fatalities [24, 25].
- **Cherry Mine Disaster.** This disaster occurred at the Cherry Mine, Illinois on November 13, 1909 and is considered one of the top ten mining disasters in the United States with 259 fatalities [26, 27].
- **Sunjiawan Mine Disaster.** This disaster occurred at theSunjiawan Mine located at Fuxin city, Liaonning province, People's Republic of China on February 14, 2005 and killed at least 214 people [28].
- **Renard Coal Mine Disaster.** This disaster occurred at the Renard Coal Mine located at Sonsnowiec, Poland in 1880 and caused 200 fatalities [29].
- **Hillcrest Mining Disaster.** This disaster occurred at the Hillcrest Coal Mine located at Hillcrest, Alberta, Canada on June 19, 1914 and killed 189 people [30, 31].
- **Nanaimo Mining Disaster.** This disaster occurred at the Nanaimo Coal Mine, located at Nanaimo, British Columbia, Canada on May 23, 1887 and caused 150 fatalities [32].

- **Ulyanovskya Coal Mine Disaster.** This disaster occurred at Ulyanovskya Coal Mine located at Novokuznetsk, Kuzbass Siberia, Russia on March 19, 2007 and caused 108 fatalities [33].
- **Mount Kembla Mining Disaster.** This disaster occurred at Mount Kembla Coal Mine in the Illawarra District of New South Wales, Australia on July 31, 1902 and killed 96 people [34, 35].
- **Mount Mulligan Mining Disaster.** This disaster occurred at the Mount Mulligan Mine located at Mount Mulligan, Queensland, Australia on September 19, 1921 and caused 75 fatalities [35, 36].
- **Springhill Mining Disaster.** This disaster occurred at the Springhill Coal Mine located at Springhill, Nova Scotia, Canada on October 23, 1958 and killed 74 people [37].

1.5 Terms and Definitions

There are many terms and definitions used in the area of safety and related areas. Some of these that are considered useful for application in the area of mine safety [2, 7, 38–41] are:

- **Mine.** This is an excavation from which ore or minerals are extracted.
- **Ore.** This is any natural combination of minerals.
- **Open-pit mining.** This is a form of operation designed to extract minerals that lie close to the earth's surface.
- **Safety.** This is the conservation of human life and the maintenance of its effectiveness, and the prevention of damage to items/systems as per specified requirements.
- **Accident.** This is an undesired and unplanned act.
- **Hazard.** This includes physical, physiological, and behavioral factors that, when not properly controlled, result in harmful incidents.
- **Unsafe condition.** This is any condition that under the right circumstances will result in an accident.
- **Safety process.** This is a set of procedures followed to enable the safety requirements of an item/system to be identified and satisfied.
- **Safety assessment.** This is the qualitative/quantitative determination of safety.
- **Safety management.** This is the accomplishment of safety through the efforts of other individuals.
- **Unsafe act.** This is an act that is not safe for an individual/worker.
- **Safeguard.** This is a barrier guard, device, or procedure developed for protecting people.
- **Hazard control.** This is a means of reducing the risk from exposure from a perceived hazard.
- **Safety plan.** This is the implementation details of how the safety requirements of the project will be satisfied.

- **Failure.** This is the inability of an item/system to operate within stated guidelines.
- **Hazard rate.** This is the ratio of the change in the number of items/units that have failed to the number of items/units that have survived at time t.
- **Human factors.** This is a study of the interrelationships between humans, the tools they use, and the surrounding environment in which they work and live.
- **Human error.** This is the failure to perform a specified task (or the performance of a forbidden action) that could lead to disruption of scheduled operations or result in damage to equipment and property.
- **Mission time.** This is the time during which an item/system is carrying out its stated function.

1.6 Useful Information on Mine Safety

There are many sources for obtaining mine safety-related information. This section categorically lists some of the most useful sources for obtaining such information, either directly or indirectly.

1.6.1 Organizations

- World Safety Organization, P.O. Box No. 1, Lalong Laan Building, Pasay City, Merto Manila, The Philippines.
- National Safety Council, 444 North Michigan Avenue, Chicago, Illinois, USA.
- American Society of Safety Engineers, 1800 East Oakton St., Des Plaines, Illinois, USA.
- System Safety Society, 14252 Culver Drive, Suite A-261, Irvine, California, USA.
- The American Institute of Mining, Metallurgical, and Petroleum Engineers, 8307 Shaffer Parkway, Littleton, Colorado, USA.
- Occupational Safety and Health Administration (OSHA), U.S. Department of Labor, 200 Constitution Avenue, Washington, D.C., USA.
- British Safety Council, 62 Chancellors Road, London, UK.
- National Institute for Occupational Safety and Health (NIOSH), 200 Independence Avenue SW, Washington DC, USA.
- Pacific Safety Council, 7220 Trade, Suite # 100, San Diego, California, USA.
- Institution of Occupational Safety and Health, The Grange Highfield Drive, Wigston, Leicestershire, UK.
- Board of Certified Safety Professionals, 208 Burwash Avenue, Savoy, Illinois, USA.
- Mines and Aggregates Safety and Health Association, 690 McKeown Avenue, North Bay, Ontario, Canada.

- Institution of Mining and Metallurgy, Danum House, South Parade, Doncaster, U.K.
- Minnesota Mine Safety Association, P.O. Box 2073, North Mankato, Minnesota, USA.
- Safety and Health Council of New Hampshire, 163 Machester Street, Suite D, Concord, New Hampshire, USA.

1.6.2 Journals and Magazines

- Australasian Mine Safety Journal
- West Australian Journal of Occupational Safety and Health
- Professional Safety
- Hazard Prevention
- International Journal of Mining, Reclamation, and Environment
- Accident Analysis and Prevention
- National Safety News
- Australian Safety News
- Safety Management Journal
- Journal of Safety Research
- Safety and Health
- European Journal of Mineral Processing and Environment Protection
- International Journal of Mining and Geological Engineering
- International Journal of Mineral Processing
- Journal of Mining Science
- Mineral Engineering
- Mining Science and Technology
- Mining Technology
- Accident Prevention
- Safety Surveyor
- Journal of Fire Safety
- International Journal of Reliability, Quality, and Safety Engineering
- Mining Magazine
- Soviet Mining Science
- Coal Age

1.6.3 Books

- Bird, F.E., Mine Safety and Loss Control: A Management Guide, Institute Press, Loganville, Georgia, 1980.
- Simpson, G., Horberry, T., Joy, J., Understanding Human Error in Mine Safety, Ashgate Publishing, London, 2009.

- Newhouse, T.V., Editor, Coal Mine Safety, Nova Science Publishers, New York, 2009.
- Gunningham, N., Mine Safety: Law Regulation Policy, The Federation Press, Annadale, New South Wales, Australia, 2007.
- Dhillon, B.S., Mining Equipment Reliability, Maintainability and Safety, Springer, London, 2008.
- Bryan, A., The Evolution of Health and Safety in Mines, Ashire Publishing, London, 1975.
- Kneeland, F.H., Weigel, W.M., Collins, H.E., Mining Equipment and Mine Organization and Safety, McGraw Hill Book Company, New York, 1915.
- Brunton, D.W., Davis, J.A., Safety and Efficiency in Mine Tunneling, Government Printing Office, Washington, D.C., 1916.
- Dhillon, B.S., Engineering Safety: Fundamentals, Techniques, and Applications, World Scientific Publishing, River Edge, New Jersey, 2003.
- Goetsch, D.L., Occupational Safety and Health, Prentice Hall, Englewood Cliffs, New Jersey, 1996.
- Hammer, W., Price, D., Occupational Safety and Engineering, Prentice Hall, Upper Saddle River, New Jersey, 2001.
- Spellman, F.R., Whiting, N.E., Safety Engineering: Principles and Practice, Government Institutes, Rockville, Maryland, 1999.
- Handley, W., Industrial Safety Handbook, McGraw Hill Book Company, London, 1969.
- Whiteside, J., Regulating Danger: The Struggle for Mine Safety in the Rocky Mountain Coal Industry, University of Nebraska Press, Lincoln, Nebraska, 1990.

1.6.4 Conference Proceedings

- Proceedings of the Annual Meeting of the Society for Mining, Metallurgy, and Exploration.
- Proceedings of the Symposium on Control of Respirable Coal Mine Dust, 1983.
- Proceedings of the National Seminar on Mine Ventilation, Safety, and Environment, 2001.
- Proceedings of the 6th Institute on Coal Mine Health and Safety, 1980.
- Proceedings of the CENTO Symposium on Mine Health and Safety, 1966.
- Proceedings of the International Conference on Radiation Hazards in Mining: Control, Measurement, and Medical Aspects, 1981.
- Proceedings of the Conference on Occupational Exposures to Fibrous and Particulate Dust and Their Extension into Environment, 1977.
- Proceedings of the 6th International Mine Ventilation Congress, 1997.
- Proceedings of the National Conference on Medicine and the Federal Coal Mine Health and Safety Act of 1969, 1970.
- Proceedings of the Annual Institute on Mining Health, Safety, and Research.
- Proceedings of the American Mining Congress-Coal Convention, 1991.

1.6.5 Data Information Sources

- Government Industry Data Exchange Program (GIDEP), GIDEP Operations Center, US Department of the Navy, Corona, California, USA.
- Safety Research Information Service, National Safety Council, 444 North Michigan Avenue, Chicago, Illinois, USA.
- International Occupational Safety and Health Information, Center Bureau, International du Travail, CH-1211, Geneva, Switzerland.
- National Technical Information Service (NTIS), 5285 Port Royal Road, Springfield, Virginia, USA.
- Computer Accident/Incident Report System, System Safety Development Center, EG8G, P.O. Box 1625, Idaho Falls, Idaho, USA.
- National Electronic Injury Surveillance System, U.S. Consumer Product Safety Commission, 5401 Westbard Street, Washington, D.C., USA.

1.7 Scope of the Book

As in other areas of the industrial sector, mining safety is important.

In recent years, mine safety has become a pressing issue because each year thousands of miners die or are injured from mining accidents throughout the world. These accidents have a variety of causes including dust explosions, collapsing of mine stopes, flooding, and general mechanical errors from improperly used or malfunctioning mining systems or equipment.

Over the years a large number of publications have appeared that are directly or indirectly related to mine safety. To the best of author's knowledge, there is at present no book whose framework includes up-to-date developments in mine safety. This book not only attempts to provide up-to-date coverage of the ongoing efforts in mine safety, but also covers useful developments in general safety.

Finally, the main objective of this book is to provide professionals concerned with safety in mines with the latest information that could contribute to the reduction or elimination of safety-related problems. This book will be useful to many individuals, including engineering and safety professionals working in the mining industry, mining administrators, mining engineering undergraduate and graduate students, researchers and instructors in the area of mining, and human factors specialists.

1.8 Problems

1. Write an essay on mine safety.
2. Compare the terms "safety" and "mine safety".
3. Define the following terms:

- Safeguard
- Accident
- Hazard

4. Discuss the need for improving safety in mines.
5. List at least eight facts and figures concerned with mine safety.
6. Discuss at least five major mine disasters.
7. Write down five sources for obtaining mine safety-related information.
8. Define the following terms:

   - Mine
   - Unsafe act
   - Unsafe condition

9. List at least seven important books for obtaining, directly or indirectly, mine-safety-related information.
10. List at least six journals considered most useful for obtaining mine safety-related information.

References

1. Mine Safety and Health Administration (MSHA), US Department of Labor, Washington, D.C. [Online]. Available from: URL: http://en.wikipedia.org/wiki/Mine_Safety_and_Health_Administration
2. Dhillon BS. Mining Equipment Reliability, Maintainability, and Safety. London: Springer; 2008.
3. 30 CFR 18.2, Code of Federal Regulations, Mine Safety and Health Administration (MSHA), US Department of Labor, Washington, D.C.
4. Division of information Services, United States Bureau of Labor Statistics, 2 Massachusetts Avenue, Washington, D.C.
5. National Institute for Occupational Safety and Health (NIOSH), 200 Independence Avenue SW, Washington, D.C.
6. Report on Injuries in America in 2000, National Safety Council, Chicago, Illinois, 2000.
7. Spellman FR, Whiting NE. Safety Engineering: Principles and Practice. Rockville, Maryland: Government Institutes; 1999.
8. All Mining Fatalities by State, Mine Safety and Health Administration, US Department of Labor, Washington, D.C. [Online]. 2007 Jan 15; Available from: URL: www.msha.gov/stats/charts/allstatesnew.asp.
9. Coal Fatalities by State, Mine Safety and Health Administration, US Department of Labor, Washington, D.C. [Online]. 2007 Jan 15; Available from: URL: www.msha.gov/stats/charts/coalbystate.asp.
10. Cawlley JC. Electrical Accidents in the Mining Industry, 1990–1999. IEEE Trans Ind Appl 2003;39(6):1570–1576.
11. Scott A. Killing Off Errors. Mine Quarry, May 1995: 14–18.
12. Historical Data on Mine Disasters in the United States, US Department of Labor, Washington, D.C. [Online]. Available from: URL: www.msha.gov/msainfo/factssheets /mshafacts8.htm.
13. MSHA Data for 1978–1988, Mine Safety and Health Administration (MSHA), US Department of Labor, Washington, D.C.
14. Hamilton DD, Hopper JE, Jones JH. Inherently Safe Mining Systems, Report No. USBM OFR 124–77, US Bureau of Mines (USBM), Washington, D.C., 1977.

15. De Rosa M. Equipment Fires Cause Injuries: NIOSH Study Reveals Trends for Equipment Fires at U.S. Coal Mines, Coal Age, October 2004: 28–31.
16. Burgess-Limerick R, Steiner L. Preventing Injuries: Analysis of Injuries Highlights High Priority Hazards Associated with Underground Coal Mining Equipment. Am Longwall Mag. August 2006: 19–20.
17. Iannacchione A, Mucho T. 100 Years of Improvement in Aggregate Worker Safety, Stone Sand Gravel Rev. March/April 2003: pp. 28–34.
18. Rethi LL, Barett EA. A Summary of Injury Data for Independent Contractor Employees in the Mining Industry from 1983–1990, Report No. USBMIC 9344, US Bureau of Mines. Washington, D.C.1983.
19. Honkeiko Colliery Mining Disaster. London: Encyclopaedia Britannica; 2009. Available from: URL: www.britannica.com/EBchecked/topic/1503377/Honkeiko-colliery-mining-disaster.
20. Brown JH. The Valley of the Shadow: An Account of Britain's Worst Mining Disaster, The Senghenydd Explosion. Port Talbot, U.K: Alwn Books; 1981.
21. Phillips JB. Senghenydd: A Brave Community. Abertillery, U.K.: The Old Bakehouse Publications; 2002.
22. Leger JP. Trends and Causes of Fatalities in South African Mines, Safety Science 1991; 14:169–185.
23. Duckham H, Duckham BF. Great Pit Disasters: Great Britain 1700 to the Present Day. London: David and Charles; 1973.
24. Monongah Mining Disaster of 1907. London: Encyclopedia Britannica; 2009. Available from: URL: www.britannica.com/EBchecked/topic/389922/Monongah-mining-disaster-of-1907.
25. Monongah Mining Disaster, Wikipedia, 2009. Available from: URL: http://en.wikipedia.org/wiki/Monongah-Mining-Disaster.
26. Tintory K. Trapped: The 1909 Cherry Mine Disaster (Illinois). Memphis, Tennessee: Atria Publishing; 2002.
27. Stout S. Story of the Great Cherry Coal Mine Disaster, Journal of the Illinois State Historical Society 1979; Vol. LXII(1):3–5.
28. Sunjiawan Mine Disaster, Wikipedia, 2009. Available from: URL: http://en.wikipedia.org/wiki/2005-Sunjiawan-mine-disaster.
29. List of Mining Disasters in Poland, Wikipedia, 2009. Available from: URL: http://en.wikipedia.org/wiki/List_of_mining_disasters_in_Poland.
30. Anderson FW. Canada's Worst Mine Disaster. Calgary, Canada: Frontier Books; 1969.
31. Hillcrest Mine Disaster, Wikipedia, 2009. Available from: URL: http://en.wikipedia.org/wiki/Hillcrest-mine-disaster.
32. 1887 Nanaimo Mine Explosion, Wikipedia, 2009. Available from: URL: http://en.wikipedia.Nanaimo-mine-explosion.
33. Ulyanovskaya Mine Disaster, Wikipedia, 2009. Available from: URL: http://en.wikipedia.org/wiki/Ulyanovskaya-Mine-disaster.
34. Piggin S, Lee H. The Mount Kembla Disaster. Melbourne, Australia: Oxford University Press; 1992.
35. Mining in Australia Wikipedia, 2009. Available from: URL: http://en.wikipedia.org/wiki/Mining-in-Australia.
36. Barwick, Australia's Worst Disasters: Mining Disasters, Heinnemann, Port Melbourne, Victoria, Australia, 1999.
37. Springhill Mining Disaster, Wikipedia, 2009. Available from: URL: http://en.wikipedia.org/wiki/Springhill_Mining_Disaster.
38. Dictionary of Terms Used in the Safety Profession, American Society of Safety Engineers, Des Plaines, Illinois, 1988.
39. McKenna T, Oliverson R. Glossary of Reliability and Maintenance Terms. Houston, Texas: Gulf Publishing; 1997.
40. MIL-STD-721, Definitions of Effectiveness Terms for Reliability, Maintainability, Human Factors, and Safety, US Department of Defense, Washington, DC.
41. Goetsch DL. Occupational Safety and Health. Englewood Cliffs, New Jersey: Prentice Hall; 1996.

42. Meulen MVD. Definitions for Hardware and Software Safety Engineers. London: Springer-Verlag; 2000.
43. Dhillon BS. Engineering Safety: Fundamentals, Techniques, Applications. River Edge, New Jersey: World Scientific Publishing; 2003.
44. Dhillon BS. Human Reliability with Human Factors. New York: Pergamon Press; 1986.

Chapter 2
Safety Mathematics and Basics

2.1 Introduction

The history of our currently used numerals can be traced back to around 250 B.C. to the stone columns erected by the Scythian emperor named Asoka of India [1], even though there is evidence that the concept of number and the process of counting developed much earlier than the time of recorded history. However, the history of the probability concept is relatively new. It goes back to the writings of Girolamo Cardano (1501–1576), in which he considered various interesting probability-related issues [1, 2]. Today, various basic mathematics and probability concepts play an important role in performing numerous types of safety analysis.

Although the history of the safety field may be traced back to ancient times, its modern development is traced to 1868 when a patent was awarded for first safeguard, and to 1877 when the Massachusetts legislature passed a law requiring adequate safeguards on hazardous machinery [3, 4]. Today, safety has become an important issue because each year thousands of people die and are seriously injured due to workplace and other accidents. For example, in 1996 in the United States there were 93,400 deaths and a large number of disabling injuries due to accidents at large [5].

This chapter presents various important aspects of safety mathematics and basics considered useful to the understanding of subsequent chapters of this book.

2.2 Arithmetic Mean, Mean Deviation, and Standard Deviation

Any data set can only be useful if it is analyzed according to certain characteristics that describe the nature of data, thus producing better decisions. This section presents three statistical measures considered useful for application in the area of mine safety [6].

2.2.1 *Arithmetic Mean*

This is defined by

$$m = \frac{\sum_{j=1}^{k} x_j}{k} \tag{2.1}$$

where

m is the arithmetic mean.
k is the total number of values.
x_j is the jth value; for $j = 1, 2, 3, ---, k$.

Example 2.1

Assume that the number of monthly accidents in a mine during a six month period were 8, 2, 20, 7, 10, and 15. Calculate the mean (i.e., arithmetic mean) number of accidents per month in the mine.

By substituting the specified data values into Equation 2.1, we obtain

$$m = \frac{8+2+20+7+10+15}{6}$$

$$= 10.33 \text{ accidents/month}$$

Thus, the mean number of accidents per month in the mine is 10.33.

2.2.2 *Mean Deviation*

This is an easily calculated measure of dispersion and is expressed by

$$D_m = \frac{\sum_{j=1}^{k} |y_j - m|}{k} \tag{2.2}$$

where

D_m is the mean deviation.
m is the mean of the given data set.
k is the number of data points in a given data set.
y_j is the jth data value, for $j = 1, 2, \ldots, k$.
$|y_j - m|$ is the absolute value of the deviation of y_j from m.

Example 2.2

Assume that safety-related monthly problems in a mine over the period of seven months were 20, 5, 15, 10, 30, 25, and 12. Calculate the mean deviation of the given data set.

By substituting the given data values into Equation 2.1, we get

$$m = \frac{20+5+15+10+30+25+12}{7}$$

$$= 16.71 \text{ safety-related problems/month}$$

Using the calculated value and the data values specified above in Equation 2.2, we get

$$\begin{aligned} D_{\mathrm{m}} &= [|20-16.71|+|5-16.71|+|15-16.71|+|10-16.71| \\ &\quad +|30-16.71|+|25-16.71|+|12-16.71|] \\ &= 7.10 \end{aligned}$$

Thus, the mean deviation of the given set of data is 7.10.

2.2.3 *Standard Deviation*

This is a commonly used measure of dispersion and is defined by

$$\sigma = \left[\frac{\sum_{j=1}^{k} (y_j - \mu)^2}{k} \right]^{1/2} \tag{2.3}$$

where

- σ is the standard deviation.
- μ is the mean.
- k is the number of data points in a given set of data.
- y_j is the jth data value; for $j = 1, 2, 3, \ldots, k$.

The standard deviation properties relating to the widely known normal distribution are as follows [6]:

- 68.27% of the values or cases are included between $\mu+\sigma$ and $\mu-\sigma$.
- 95.45% of the values or cases are included between $\mu+2\sigma$ and $\mu-2\sigma$.
- 99.73% of the values or cases are included between $\mu+3\sigma$ and $\mu-3\sigma$.

Example 2.3

Calculate the standard deviation of the data set given in Example 2.2.

From Example 2.2, we get

$$\mu = m = 16.71 \text{ safety-related problems/month}$$

$\mu = m = 16.71$ safety-related problems/month

Using the above calculated value and the specified data values in Equation 2.3 yields

$$\begin{aligned}\sigma &= [\{(20-16.71)^2 + (5-16.71)^2 + (15-16.71)^2 + (10-16.71)^2 + (30-16.71)^2 \\ &\quad + (25-16.71)^2 + (12-16.71)^2\}/7]^{1/2} \\ &= 8.13\end{aligned}$$

Thus, standard deviation of the data set is 8.13.

2.3 Boolean Algebra Laws and Probability Definition and Properties

As Boolean algebra and probability theory play an important role in performing various types of analysis in safety engineering and related areas, this section presents Boolean algebra laws, probability definition, and probability properties separately [7–11].

2.3.1 Boolean Algebra Laws

Some of the Boolean algebra laws considered useful to perform various types of mine safety-related analysis are presented below:

$$X + Y = Y + X \tag{2.4}$$

where

- Y is an arbitrary set/event.
- X is an arbitrary set/event.
- $+$ denotes the union of sets.

$$X.Y = Y.X \tag{2.5}$$

where

Dot (.) denotes the intersection of sets. It is to be noted that when Equation 2.5 or any other Boolean algebra equation is written without the dot, it still conveys the same meaning.

$$X(Y+Z) = XY + XZ \tag{2.6}$$

where

Z is the arbitrary set/event.

$$XX = X \tag{2.7}$$

$$X + X = X \tag{2.8}$$

$$X(X+Y) = X \tag{2.9}$$

$$X + XY = X \tag{2.10}$$

$$(X+Y)+Z = X+(Y+Z) \tag{2.11}$$

$$(XY)Z = X(YZ) \tag{2.12}$$

$$X + YZ = (X+Y)(X+Z) \tag{2.13}$$

2.3.2 Probability Definition

Probability is expressed by [11]

$$P(Y) = \lim_{n\to\infty}(N/n) \tag{2.14}$$

where

N denotes the number of times event Y occurs in the n repeated experiments.

$P(Y)$ is the probability of occurrence of event Y.

2.3.3 Probability Properties

Some of the basic probability properties are as follows [9, 11]:

- The probability of occurrence of event, say X, is

$$0 \le P(X) \le 1 \tag{2.15}$$

- Probability of the sample space S is

$$P(S) = 1 \tag{2.16}$$

- Probability of the negation of the sample S is

$$P(\overline{S}) = 0 \tag{2.17}$$

where

$\overline{S}$ is the negation of the sample S.

- The probability of occurrence and non-occurrence of an event, say Y, is

$$P(Y) + P(\overline{Y}) = 1 \tag{2.18}$$

where

$P(Y)$	is the probability of occurrence of event Y.
$P(\overline{Y})$	is the probability of non-occurrence of event Y.

- The probability of the union of m independent events is

$$P(Y_1 + Y_2 + --- + Y_m) = 1 - \prod_{j=1}^{m} (1 - P(Y_j)) \tag{2.19}$$

where

Y_j	is the jth event; for $j = 1, 2, \ldots, m$.
$P(Y_j)$	is the probability of occurrence of event Y_j; for $j = 1, 2, \ldots, m$.

- The probability of the union of m mutually exclusive events is

$$P(Y_1 + Y_2 + --- + Y_m) = \sum_{j=1}^{m} P(Y_j) \tag{2.20}$$

- The probability of an intersection of m independent events is

$$P(Y_1 Y_2 --- Y_m) = P(Y_1)P(Y_2) --- P(Y_m) \tag{2.21}$$

Example 2.4

A mining safety system is composed of three critical subsystems Y_1, Y_2, and Y_3. The failure of any one of these subsystems can cause an accident. The failure probabilities of subsystems Y_1, Y_2, and Y_3 are 0.05, 0.15, and 0.1, respectively. Find the probability of the occurrence of an accident if all the three subsystems fail independently.

By substituting the given data values into Equation 2.19, we get

$$\begin{aligned} P(Y_1 + Y_2 + Y_3) &= 1 - \prod_{j=1}^{3} (1 - P(Y_j)) \\ &= 1 - [1 - P(Y_1)][1 - P(Y_2)][1 - P(Y_3)] \\ &= 1 - [1 - 0.05][1 - 0.15][1 - 0.1] \\ &= 0.2732 \end{aligned}$$

Thus, the probability of the occurrence of an accident is 0.2732.

2.4 Probability Distributions

Usually, various types of probability distributions are used to perform mathematical safety analysis. Some of the common ones used to perform mine safety mathematical analysis are presented below [12, 13].

2.4.1 Exponential Distribution

This is one of the simplest probability distributions used to conduct various types of safety analysis. The probability density of the distribution is expressed by [14]

$$f(t) = \lambda e^{-\lambda t}, \text{ for } \lambda > 0, t \geq 0 \tag{2.22}$$

where

- t is time.
- $f(t)$ is the probability density function.
- λ is the distribution parameter. In safety studies, it is often referred to as constant unsafe or safe occurrence rate.

For continuous random variables, the cumulative distribution function is expressed by [12–14]

$$F(t) = \int_{-\infty}^{t} f(x)\mathrm{d}x \tag{2.23}$$

where

- $F(t)$ is the cumulative distribution function.

Thus, by inserting Equation 2.22 into Equation 2.23, we obtain the following expression for the exponential cumulative distribution function:

$$F(t) = 1 - \mathrm{e}^{-\lambda t} \tag{2.24}$$

Example 2.5

In a coal mine, the monthly accident occurrence rate is 0.06 accidents per month. Calculate the probability of an accident occurrence during a 12-month period, if the accident occurrence times are exponentially distributed.

We have

$$\lambda = 0.06 \text{ accidents/month}$$

and

$$t = 12 \text{ months}$$

By substituting the above given values into Equation 2.24, we get

$$F(12) = 1 - e^{-(0.06)(12)}$$
$$= 0.5132$$

Thus, the probability of an accident occurrence during the 12-month period is 0.5132.

2.4.2 *Normal Distribution*

This is a well known probability distribution that was discovered by De Moivre in 1733 [15]. The probability density function of the distribution is defined by

$$f(t) = \frac{1}{\sigma\sqrt{2\pi}} \exp\left[-\frac{(t-\mu)^2}{2\sigma^2}\right], -\infty \langle t \langle +\infty \tag{2.25}$$

where

t is time.
σ and μ are the distribution parameters (i.e., standard deviation and mean, respectively).

By substituting Equation 2.25 into Equation 2.23, we get the following equation for the cumulative distribution function:

$$F(t) = \frac{1}{\sigma\sqrt{2\pi}} \int_{-\infty}^{t} \exp\left[-\frac{(x-\mu)^2}{2\sigma^2}\right] dx \tag{2.26}$$

2.4.3 *Weibull Distribution*

This probability distribution was developed by W. Weibull and its probability density function is defined by [16]

$$f(t) = \frac{\theta t^{\theta-1}}{\alpha^\theta} e^{-(t/\alpha)^\theta}, \theta\rangle 0, \alpha\rangle 0, t \geq 0 \tag{2.27}$$

where

t is time.
θ is the distribution shape parameter.
α is the distribution scale parameter.

By substituting Equation 2.27 into Equation 2.23, we obtain the following equation for the distribution cumulative distribution function:

$$F(t) = 1 - e^{-(t/\alpha)^{\theta}} \tag{2.28}$$

It is to be noted that for $\theta = 1$ in Equations 2.27 and 2.28 the Weibull distribution reduces to an exponential distribution. More specifically, for $\theta = 1$ and $\alpha = \frac{1}{\lambda}$, Equations 2.27 and 2.28 are identical to Equations 2.22 and 2.24, respectively.

2.5 Expected Value and Laplace Transform Definitions and Final Value Theorem

The definitions of expected value and Laplace transforms and the final value theorem are separately presented below.

2.5.1 *Expected Value*

The expected value of a continuous random variable is defined by [11, 15]

$$E(t) = \mu = \int_{-\infty}^{\infty} t f(t) \, dt \tag{2.29}$$

where

t is a continuous random variable.
$E(t)$ is the expected value of t.
μ is the mean value.
$f(t)$ is the probability density function.

2.5.2 *Laplace Transform*

The Laplace transform of the function $f(t)$ is defined by

$$f(s) = \int_{0}^{\infty} f(t) e^{-st} \, dt \tag{2.30}$$

where
t is the time variable.
s is the Laplace transform variable.
$f(t)$ is a function.
$f(s)$ is the Laplace transform of $f(t)$.

Example 2.6

Obtain the Laplace transform of the following function:

$$f(t) = \mathrm{e}^{\theta t} \tag{2.31}$$

where

t is time.
θ is a constant.

By substituting Equation 2.31 into Equation 2.30, we get

$$\begin{aligned} f(s) &= \int_0^\infty \mathrm{e}^{\theta t}\mathrm{e}^{-st}\,\mathrm{d}t \\ &= \int_0^\infty \mathrm{e}^{-(s-\theta)t}\,\mathrm{d}t \\ &= -\frac{\mathrm{e}^{-(s-\theta)t}}{(s-\theta)}\Bigg|_0^\infty \\ &= \frac{1}{(s-\theta)} \end{aligned} \tag{2.32}$$

Table 2.1 presents Laplace transforms of functions considered useful to perform mine safety-related mathematical analysis [17, 18].

Table 2.1 Laplace transforms of some functions

$f(t)$	$f(s)$
c(a constant)	c/s
t^m, *for* $m = 0, 1, 2, 3, \ldots$	$m!/s^{m+1}$
$t\mathrm{e}^{-ct}$	$1/(s+c)^2$
e^{-ct}	$1/(s+c)$
$c_1 f_1(t) + c_2 f_2(t)$	$c_1 f_1(s) + c_2 f_2(s)$
$\dfrac{\mathrm{d}f(t)}{\mathrm{d}t}$	$sf(s) - f(0)$

2.5.3 *Laplace Transform: Final Value Theorem*

If the following limits exist, then the final value theorem is defined by

$$\lim_{t\to\infty} f(t) = \lim_{s\to 0}\left[sf(s)\right] \tag{2.33}$$

Example 2.7

Prove by using the following equation that the left side of Equation 2.33 is equal to its right side:

$$f(t) = \mathrm{e}^{-\lambda t} \tag{2.34}$$

where

λ is a constant.
t is time.

By substituting Equation 2.34 into the left side of Equation 2.33, we get

$$\lim_{t\to\infty} \mathrm{e}^{-\lambda t} = 0$$

By substituting Equation 2.34 into Equation 2.30, we obtain

$$\begin{aligned} f(s) &= \int_0^\infty \mathrm{e}^{-(s+\lambda)t}\,\mathrm{d}t \\ &= \frac{1}{(s+\lambda)} \end{aligned} \tag{2.35}$$

By substituting Equation 2.35 into the right side of Equation 2.33, we obtain

$$\lim_{s\to 0}\left[\frac{s}{(s+\lambda)}\right] = 0$$

Thus, the above results prove that the left side of Equation 2.33 is equal to its right side.

2.6 Solving First Order Differential Equations Using Laplace Transforms

The application of the Markov method to mathematical mine safety analysis sometimes results in a system of linear first-order differential equations. These equations must be solved to find state probabilities at a given time t. Although there are various methods that can be used for this purpose, the Laplace transform method appears to be the most effective.

The application of Laplace transforms to find solutions to a set of first order linear differential equations describing a simple mining system is demonstrated through the example presented below.

Example 2.8

Assume that a mining system is either operating normally or has failed. The following two first order linear differential equations describe such a system:

$$\frac{\mathrm{d}P_0(t)}{\mathrm{d}t} + \lambda_m P_0(t) = 0 \tag{2.36}$$

$$\frac{\mathrm{d}P_f(t)}{\mathrm{d}t} - \lambda_m P_0(t) = 0 \tag{2.37}$$

where

λ_m is the mining system constant failure rate.
$P_j(t)$ is the probability that the mining system is in state j at time t; for $j=0$ (operating normally), $j=f$ (failed).

At time $t=0$, $P_0(0)=1$ and $P_f(0)=0$.

Find solutions to Equations 2.36 and 2.37 by using the Laplace transform method.

By taking Laplace transforms of Equations 2.36 and 2.37 and using Table 2.1 and the given initial conditions, we obtain

$$P_0(s) = \frac{1}{s+\lambda_m} \tag{2.38}$$

$$P_f(s) = \frac{\lambda_m}{s(s+\lambda_m)} \tag{2.39}$$

By taking the inverse Laplace transforms of Equations 2.38 and 2.39, we get

$$P_0(t) = e^{-\lambda_m t} \tag{2.40}$$

$$P_f(t) = 1 - e^{-\lambda_m t} \tag{2.41}$$

Thus, Equations 2.40 and 2.41 are the solutions to differential Equations 2.36 and 2.37.

2.7 Safety and Engineers

The problem with the safety of engineering products is not new and may be traced back to the early railroads. For example, on the day of Stephenson's first railroad line dedication, a railroad accident killed a prominent English legislator [19]. The following year, the boiler of the first locomotive built in the United States exploded and caused one death and badly injured a number of people [19, 20].

Today, engineering products and systems have become highly complex and sophisticated. The safety of these products and systems has become a challenging issue, and because of competitive financial environments and other factors engineers are undergoing significant pressure to rapidly complete new designs at minimum costs. Experience has shown that this often leads to more design-related deficiencies and errors which can result in accidents.

Design deficiencies may result from factors such as those listed below [19]:

- Design is incomplete, confusing, or wrong
- Design relies on product users to avoid accidents
- Designer overlooked elimination or reduction of the possibility of human errors
- Designer overlooked effective warning of potential hazards
- Design violates usual tendencies/capabilities of potential users
- Design does not properly determine or consider the consequences of failure, error, action, or omission
- Designer overlooked an unexpected application of a product or its potential consequences
- Design incorporates poor warning mechanisms
- Design places an unreasonable level of stress on product operators
- Designer overlooked prescription of appropriate operational procedures in conditions where hazards might exist

2.8 Statute, Administrative, Common, and Liability Laws

In any jurisdiction, statute laws are promulgated by the highest governing authority or royal edict. The history of statute laws may be traced back to ancient times to the code of Hammurabi (Circa 2000 B.C.). Hammurabi was a Babylonian ruler who developed a set of rules that contained clauses concerning such issues as monetary damages assessed against those individuals who caused injury to others and allowable fees for physicians [3, 21, 22]. According to past records, an individual who caused injury to other people was treated as a criminal under early statute laws [19, 20].

Administrative laws are established or developed by an executive via prescribing the criteria under which any desired statute/control will be carried out. Past experience shows that safety-related cases do sometimes fall under the jurisdiction of administrative laws.

Common laws were first introduced in the United Kingdom and were not established by statutes but followed precedents set forth by previous judicial decisions. As can be seen from past experiences, many safety-related cases also fall under these laws.

The basic objective of liability laws is to reduce employer immunity from liability for accidents by eradicating the common law concept of assumption of risk. In the United States, the state of Alabama was the first state to pass an employer's liability law in 1885. The United States Congress passed the Occupational Safety and Health Act (OSHA) in 1970. Over the years this Act has helped to improve the safety of workplaces quite dramatically throughout the United States, as evidenced by the statistical data [22].

2.9 Accident Causation Theories

There are many accident causation theories. Some of these are as follows [3]:

- The human factors theory
- The domino theory
- The systems theory
- The combination theory
- The epidemiological theory
- The accident/incident theory

The human factors theory is based on the assumption that accidents occur due to a chain of events caused by human error, and that there are three main factors that cause human errors: overload, inappropriate activities, and inappropriate response [3, 23]. The domino theory is operationalized in ten statements called safety axioms and it assumes that there are the following five factors in the sequence of events leading up to an accident [3, 24]:

- Ancestry/social environment
- Fault of person
- Unsafe act/mechanical or physical hazard
- Accident
- Injury

The system theory is based on the assumption that any condition in which an accident might occur is a system with three elements: environment, human or person (host), and machine (agency). The combination theory posits that a single model theory cannot describe all types of accidents, as their actual causes may combine elements of various different models.

The epidemiological theory assumes that the models used to study the relationships between disease and environmental factors can also be used for determining causal relationships between environmental factors and accidents. Finally, the accident/incident theory is essentially an extension of the human factors theory and it introduces various new elements including ergonomic-related traps, the decision to err, and systems failures.

2.10 Common Causes of Work Injuries, Accident Death Rates by Industry, and Workers' Compensation

There are many causes of work-related injuries, and some of the common ones are shown in Figure 2.1 [3]. It should be noted, however, that in the United States approximately 31% of all work-related injuries are due to overexertion, according to a study performed by the National Safety Council (NSC) [3].

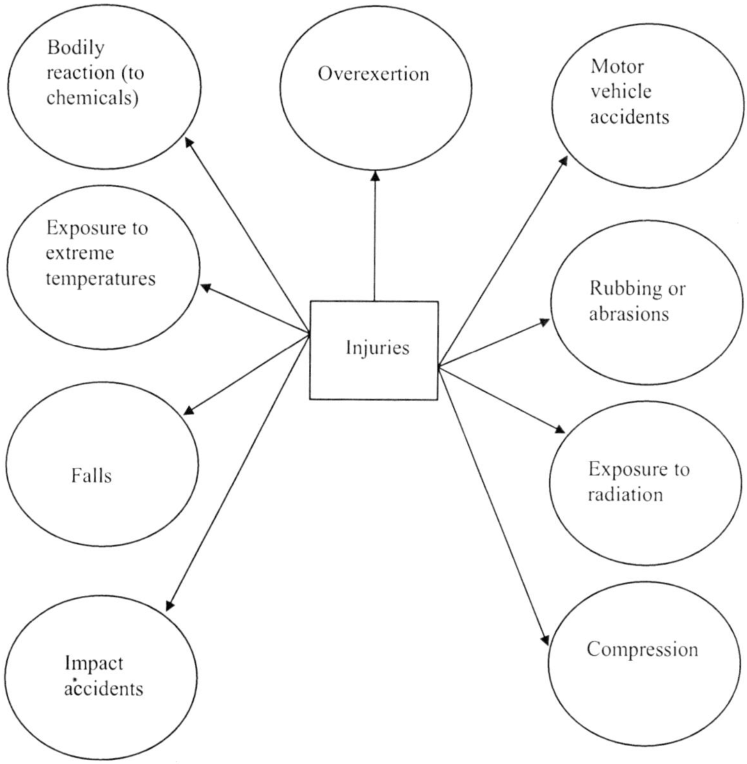

Figure 2.1 Common work-related injuries

As data on accident death rates is used for various purposes, there are many organizations and agencies in the United States that collect data on accident death rates in the industrial sector. Some of these organizations and agencies are the National Safety Council, the National Center for Health Statistics, and the Bureau of Labour Statistics. These organizations and agencies usually collect accident death rate data under eight industrial categories: manufacturing, construction, mining/quarrying, trade, agriculture, transportation/public utilities, services, and government (i.e., local, state, and federal) [3].

A ranking of accident death rates by these industrial categories for a typical year, from highest accident death rate to lowest, is presented below [3]:

- Mining/quarrying (highest accident death rate)
- Agriculture
- Construction
- Transportation/public utilities
- Government (i.e., federal, state, and local)
- Manufacturing
- Services
- Trade (lowest accident death rate)

Due to government legislation, workers' compensation has become an important element in workplace safety. Although the history of workers' compensation law may be traced back to 1838 in Prussia where it was passed for protecting railroad workers, in the United States the first workers' compensation law was enacted only in 1908 to protect federal government employees carrying out hazardous tasks [19].

In 1983, the United States Chamber of Commerce performed a study of Workers' Compensation Laws and concluded that they have seven underlying objectives [4, 19]:

- To reduce all types of human-related sufferings and preventable accidents
- To provide fast and reasonable income and medical benefits to victims of work-related accidents or income benefits to the dependents of victims, irrespective of faults
- To encourage frank study of all types of accident causes
- To provide a good mechanism for minimizing the degree of personal injury litigation in courts
- To relieve public/private charities of the financial pressure generated by uncompensated workplace-related accidents
- To eradicate time-consuming and costly trials and appeals, thus saving in the fee payments to lawyers involved
- To maximize employer involvement in safety-related matters and rehabilitation through the use of an experience-rating mechanism

Additionally, the following three basic conditions must be fully satisfied in order for an injured worker or his/her dependents to receive any benefits [19]:

- The accident in question occurred during the course of the employment.
- The injury is caused by an accident.
- The accident arose from the employment of the worker.

Finally, it is added that all costs associated with workers' compensations are born by employers as an element of their overheads.

2.11 Problems

1. What is an arithmetic mean?
2. Assume that safety-related monthly problems in a coal mine over the 5-month period were 7, 15, 4, 9, and 3, respectively. Calculate the mean deviation.
3. Mathematically define probability.
4. A mining safety system is composed of two critical subsystems X_1 and X_2. The failure of any one of these subsystems can cause an accident. The failure probabilities of subsystems X_1 and X_2 are 0.2 and 0.1, respectively. Calculate the probability of the occurrence of an accident if both the subsystems fail independently.

5. Assume that in a mine, the monthly accident occurrence rate is 0.005 accidents per month. Calculate the probability of an accident occurring during a 6-month period if the times of accident occurrence are exponentially distributed.
6. Write down Weibull distribution probability density function.
7. Take the Laplace transform of the following function:

$$f(t) = te^{-\lambda t} \tag{2.42}$$

where

t is time.
λ is a constant.

8. Write down at least ten factors that can lead to design deficiency.
9. Discuss statute and administrative laws.
10. Describe at least five accident causation theories.

References

1. Eve, H. An Introduction to History of Mathematics. New York: Holt, Rinehart, Winston; 1976.
2. Owen DB, editor. History of Statistics and Probability. New York: Marcel Dekker; 1976.
3. Goetsch DL. Occupational Safety and Health. Englewood Cliffs, New Jersey: Prentice Hall; 1996.
4. Dhillon BS. Engineering Safety: Fundamentals, Techniques, and Application. New Jersey: World Scientific Publishing; 2003.
5. Accident Facts, Report. National Safety Council. Chicago, Illinois; 1996.
6. Spiegel MR. Statistics. New York: McGraw Hill Book Company; 1961.
7. Roland HE, Moriarty B. System Safety Engineering and Management. New York: John Wiley and Sons; 1983.
8. Lipschutz S. Set Theory. New York: McGraw Hill Book Company; 1964.
9. Lipschutz S. Probability. New York: McGraw Hill Book Company; 1965.
10. Fault Tree Handbook. Report No. NUREG-0492, U.S. Nuclear Regulatory Commission. Washington, D.C.; 1981.
11. Mann NR, Schafer RE, Singpurwalla ND, Methods for Statistical Analysis of Reliability and Life Data. New York: John Wiley and Sons; 1974.
12. Patel JK, Kapadia CH, Owen DB. Handbook of Statistical Distributions. New York: Marcel Dekker, Inc.; 1976.
13. Ireson WG, editor. Reliability Handbook. New York: McGraw Hill Book Company; 1966.
14. Davis DJ. An Analysis of Some Failure Data. J Am Stat Assoc 1952; June:113–150.
15. Ramakumar R. Engineering Reliability: Fundamentals and Applications. Englewood Cliffs, New Jersey: Prentice Hall; 1993.
16. Weibull W. A Statistical Distribution Function of Wide Applicability. J Appl Mech 1951;18:293–297.
17. Spiegel MR. Laplace Transforms. New York: McGraw Hill Book Company; 1965.
18. Oberhettinger F, Badii L. Tables of Laplace Transforms. New York: Springer Verlag; 1973.
19. Hammer W, Price D. Occupational Safety Management and Engineering. Upper Saddle River, New Jersey: Prentice Hall; 2001.

20. Operator Safety. Engineering May 1974;358–363.
21. Ladon J, editor. Introduction to Occupational Health and Safety. National Safety Council (NSC). Chicago, Illinois; 1986.
22. Dhillon BS. Reliability, Quality, and Safety for Engineers. Boca Raton, Florida: CRC Press; 2005.
23. Heinrich HW, Petersen D, Roos N. Industrial Accident Prevention. New York: McGraw Hill Book Company; 1980.
24. Heinrich HW. Industrial Accident Prevention. New York: McGraw Hill Book Company; 1959.

Chapter 3
Safety Management

3.1 Introduction

Safety is an acknowledged management responsibility and its management is an important element of a safety program within an organization. The fundamental objective of safety management is to eradicate human-related suffering and anguish as well as to achieve economy of operations effectively.

The actual starting point of safety management may be regarded as the period of the 1950s and 1960s [1]. An important milestone in the history of safety management took place in 1970 when the United States Congress passed the Occupational Safety and Health Act (OSHA) [2]. Over the years, this legislation has played a pivotal role in making safety management an important element of safety programs throughout the industrial sector.

Since 1970 many new developments both directly and indirectly related to safety management have occurred. Extensive lists of publications on safety are available in [3, 4]. These include safety management, which may be simply defined as the accomplishments of safety through the efforts of others [5–7].

This chapter presents various important aspects of safety management considered useful for the mining sector.

3.2 Safety Management Principles and Safety Department Functions

There are many principles of safety management. Some of them are as follows [2, 8]:

- The safety system should be tailored to the organizational culture.
- Safety should be managed, just like managing any other activity in an organization/company. More specifically, management should direct safety efforts by establishing attainable goals, and by organizing, planning, and controlling to successfully achieve such goals.

- There is no single method for effectively achieving safety in an organization. However, in order for a safety system to be effective, it must meet certain criteria: it must be flexible, involve worker participation, have the top management clearly and visibly showing its support, etc.
- The key to effective line safety performance is management procedures that clearly factor in accountability.
- In developing an effective safety system, the three important subsystems that must be considered with care are the managerial, the behavioral, and the physical.
- All causes leading to unsafe behavior can be controlled, identified, and classified.
- There are certain groups of circumstances that can be predicted lead to severe injuries: high energy sources, unusual and non-routine activities, certain construction conditions, and non-productive activities.
- The main function of safety is to find and define the operational errors that result in accidents.

A safety department performs various types of functions. Some of the typical ones are shown in Figure 3.1 [5, 9].

3.3 Safety Manager's and Engineer's Functions

A safety manager performs many functions. Some of the important ones are as follows [5, 10]:

- Formulating and administering the safety program
- Supervising safety department employees
- Advertising safety-related issues at all management levels
- Acquiring up-to-date and best hazard control information
- Reporting to higher level management on a periodic basis concerning the state of the company's ongoing safety efforts
- Representing management to government agencies, insurance companies, public employees, etc. in regard to safety
- Directing the inspection of the company facilities with respect to compliance with the safety-related rules and regulations of outside agencies
- Participating in procurement specification reviews and in the design of new facilities/process layouts/equipment layouts
- Directing the recording and collection of useful information on items such as work-related injuries and accidents

There are many functions performed by a safety engineer. Some of the typical ones are conducting accident investigations, coordinating with management on matters concerning safety, performing safety-related studies, conducting safety inspections, collaborating with safety committees, insuring that the proper corrective actions are taken to avoid accident reoccurrences, providing appropriate safety training, and processing workers' compensation claims [9].

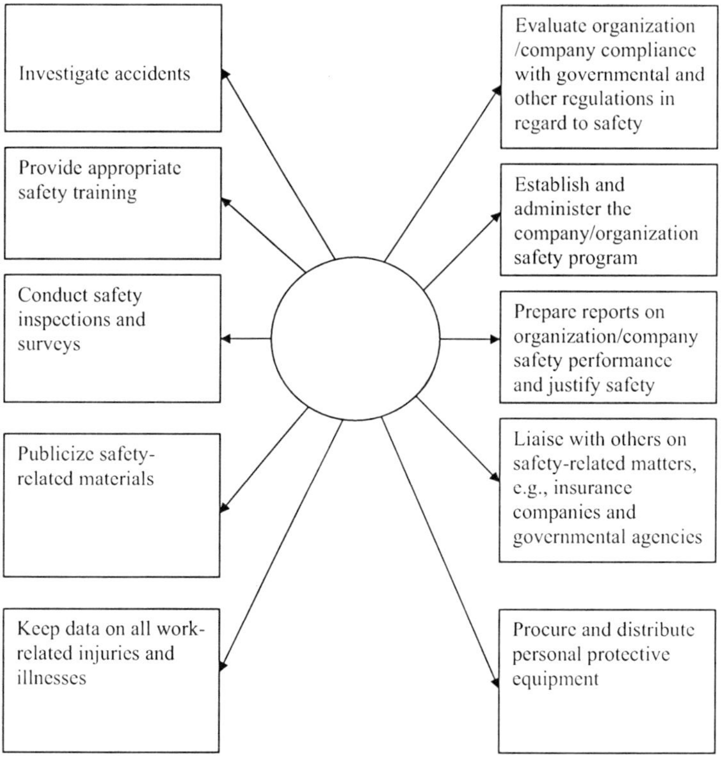

Figure 3.1 A safety department's typical functions

3.4 Developing a Safety Program Plan and Safety-related Strategies for Safety Professionals

Any company or organization contemplating introducing a safety program can develop its plan in seven steps. These steps are as follows [5, 9]:

- **Step 1:** Prepare and announce the safety policy.
  This step is concerned with announcing and writing the policy regarding controlling hazards within the organizational setup and designating accountability and authority for its implementation.
- **Step 2:** Appoint a safety chief.
  This step is basically concerned with appointing an individual with appropriate qualifications and experience to oversee safety-related matters.
- **Step 3:** Analyze operational injury records.
  This step is concerned with conducting analysis of the operational records of injuries, property damage, and work-related illnesses.

Figure 3.2 Strategies for safety professionals to reduce unsafe acts and increase safe acts

- **Step 4:** Evaluate the scope and seriousness of operational hazards.
  This step is concerned with evaluating the scope and seriousness of operational hazards. More specifically, the step is concerned with determining the nature and severity of inherent operating hazards, time estimates and budgets to carry out the corrective measures, the quality of the existing physical safeguards, etc.
- **Step 5:** Select and schedule communication methods.
  This step is concerned with selecting and scheduling communication approaches for purposes such as informing general management about the organization's safety progress and associated needs, safety-related training of workers, and interest maintenance.
- **Step 6:** Establish a schedule for periodic reviews.
  This step is concerned with establishing a schedule for periodically reviewing the program and facilities.
- **Step 7:** Develop short-range and long-range objectives.
  This step is concerned with developing appropriate short-range and long-range objectives for the safety program.

There are basically two good strategies for safety professionals to reduce the occurrence of unsafe acts and increase safe acts as shown in Figure 3.2 [2, 10]. Some of the factors associated with the strategy "Increase working safely-associated satisfactions" are as follows [2, 10]:

- Clearly emphasize the personal gains that workers reap when they carry out their tasks safely.
- Develop appropriate operational procedures/approaches not only to identify but also to reward safe practices.
- Regularly audit the occurrences of all safe practices.
- Give all involved workers tangible recognitions/awards for behaving safely.

- Make use of both enhanced individual participation and work-group participation for developing recommended safe procedures.
- Emphasize the job-related gains when working safely.
- Recognize that enforcing safe acts should be made quite distinct from discipline.
- Aim to improve the ratio of recorded commendations to recorded reprimands.
- Aim to improve the ratio of time spent in recognizing safe worker behavior in relation to the time spent to discipline workers for unsafe acts.

Similarly, the main factors associated with the strategy "Eradicate working safely- discouraging obstacles" include evaluating obstacles, discussing all proposed changes with the workers involved in order to obtain their agreement before making the change, reducing the obstacles, and examining each safety improvement on a cost-benefit basis.

3.5 Motivating Workers to Work Safely and Management-related Deficiencies Leading to Accidents

The supervisors of workers who typically recognize deviations from safe practices do their utmost best to eliminate such deviations. In this regard, there are basically two methods the supervisors pursue: eradicate unsafe acts and bolster probabilities for safe acts.

Past experience clearly indicates that in the long run the bolstering of probabilities for safe acts is a very powerful motivational exercise. Some of the ways that supervisors can benefit from this exercise/approach are as follows [2, 10]:

- Focusing attention on the importance of safe performance
- Strengthening and enhancing the importance of management standards for safe performance
- Reminding workers about the approaches/methods of safe performance
- Spending more time recognizing and rewarding safe performance than spending time disciplining workers for unsafe performance

Past experience clearly indicates that many accidents in the industrial sector occur because of management-related deficiencies. According to Fortune magazine, however, many executives in the industrial sector still believe that careless workers are really to be blamed for workplace accidents [11]. Despite this widespread belief, a survey of industrial injuries in the state of Pennsylvania showed that only 26% of the injuries were the result of workers' carelessness [12].

Some immediate causes for the occurrence of various industrial accidents are stated in the following, along with their corresponding possible management deficiencies in parentheses. These include failure to follow prescribed procedures correctly (poor enforcement to follow correct procedures, poor supervisory safety indoctrination), lack of proper procedures (poor supervisory proficiency, poor operational procedures, poor planning, layout, and design), poor housekeeping (poor supervisory training, poor planning and layout), lack of proper equipment, tools, and facilities (poor supervisory safety indoctrination, poor planning, layout, and design), defective or unsafe facilities and equipment (poor supervisory safety indoctrination, poor maintenance and repair systems, poor employee safety indoctrination), improper use of equipment, tools, and facilities (poor training of workers, poorly established operational procedures), lack of awareness of hazards involved (poor worker training, poor worker safety consciousness, poor safety rules and measures), and jobs not well understood (poorly written operational procedures, poor employee selection and placement) [13].

3.6 Safety-related Responsibilities of Non-safety Groups

In order to insure the success of a safety program within a company, the responsibility for safety rests not only with the safety group but also with other groups such as those shown in Figure 3.3 [12].

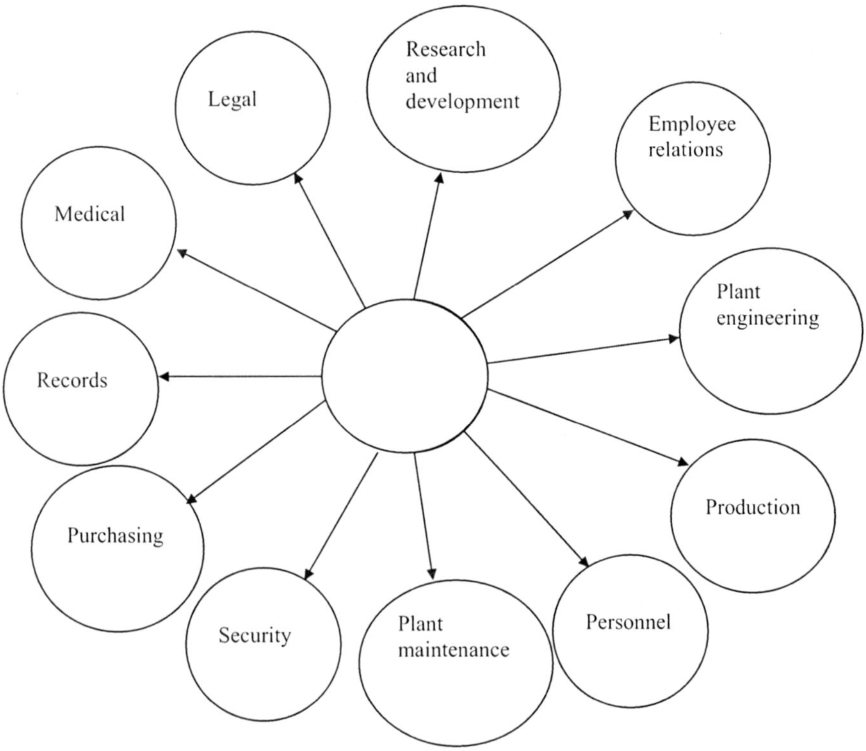

Figure 3.3 Non-safety groups responsible for insuring the success of a safety program within an organization

The plant engineering group insures that no item or equipment is selected or installed that could adversely affect the safety of all concerned individuals, unless all potential hazards can satisfactorily be safeguarded. The main responsibility of the production group is to insure with care that unsafe practices are not permitted whatsoever, even at the cost of increment in output.

The research and development group insures that whenever new products involve the testing or use of materials, any related hazards are brought to the attention of management immediately. The plant maintenance group plays an important role to insure that good housekeeping practices are maintained at all times.

The employee relations group insures factors such as follows in regard to safety:

- Maintenance of employer-union worker relationships at the highest level possible
- Immediate evaluation of worker complaints and suggestions with respect to safety
- Providing appropriate help to employees in compensation-related insurance claims
- Continuous promotion of safety-related practices on bulletin boards, in company newspapers, and through other possible means

The main objective of the purchasing group is to insure that safety-related equipment and materials are procured promptly, and that items that could be unsafe are purchased using appropriate specifications with applicable safety-related requirements. The legal group keeps all managerial personnel up-to-date on the latest safety laws as well as on judicial interpretations of laws.

The medical group insures several safety-related areas, such as effective screening of all prospective workers with regard to their mental or physical capabilities to carry out given work tasks, and effectively providing first aid and emergency treatments to all concerned individuals. The record group ensures that all information with respect to safety are analyzed, recorded, and collated for determining patterns in accident occurrences.

The main objective of the personnel group with respect to safety is to insure that all concerned workers are properly trained and physically capable of effectively carrying out their assigned tasks. Finally, the security group insures items such as:

- Unblocked emergency accesses
- Use of all company equipment by authorized personnel only
- Operating company vehicles within safe speed limits and with due care

Furthermore, the plant security group personnel can greatly assist the safety group by reporting hazards detected during their duty periods. Some examples of these hazards are missing fire extinguishers, poorly lit stairs, opened or blocked fire doors, inoperative exit lights, loose tools, metal, wires, or some other items, chemical, fuel, oil, water, or steam leaks, blocked accesses to emergency equipment, evidence of smoking in non-smoking areas, blocked emergency exists, unusual fumes or odors, damaged electrical outlets, fittings, plugs, or others devices, unlocked doors/gates enclosing hazardous areas, oily, wet, or heavily waxed floors, broken or loose hand or guard rails, obstructed lanes to be used by emergency vehicles, broken glass or uneven surfaces, poorly secured or stored gas cylinders, uncovered containers composed of fuels, solvent, or other flammable liquids, and missing barriers, lights or other protective devices at locations such as manholes, ditches, and excavations [12].

3.7 Safety Checklist for Management

Over the years various types of safety checklists have been developed for management for reviewing different aspects of plant safety as well as for determining if its associated actions are appropriate and effective. These checklists contain common questions such as follows [12]:

- Does the manager have sufficient authority to effectively carry out assigned safety-related functions?
- Does the policy directive clearly describe functions and responsibilities of each and every concerned organization in regard to safety?
- Is an effective mechanism in place to produce hazard reports?
- Does the existing safety committee meet on a regular basis to review safety matters?
- Is there an appropriate means for measuring progress of the workplace safety effort?
- Did the chief executive issue a directive identifying his/her policy toward employee safety?
- Are the policy and the safety-conduct associated rules easy to comprehend and posted at appropriate locations?
- Does the budget incorporate safety-related activities as normal functions and are these activities adequately funded?
- Is the individual supervising safety-related activities well experienced and knowledgeable in safety issues?
- Is there adequate information provided on safety matters and accident prevention to workers being trained to operate newly purchased equipment?
- Have the necessary training courses been developed for workers on safety issues?
- Is the necessary documentation concerning accidents that have occurred at the company facilities up-to-date, and is it readily available for use by Occupational Safety and Health Act (OSHA) personnel?
- Is the group involved with safety matters diverse and large enough for handling any safety problem?
- Is there an appropriate company safety committee?

- Does the chief executive have adequate knowledge of the main restrictions on health and safety imposed by OSHA?
- Do the equipment procuring documents have provisions for notifying the company's safety personnel of any potential hazard in the equipment?
- Are the rules of conduct for insuring safe conduct of the facility prepared and are they useful?
- Does the company executive's directive clearly designate a senior manager for supervising workplace safety?
- Are there appropriate mechanisms to circulate documents concerning workplace safety to all concerned people and then to store them properly?
- Is the Chief executive informed of safety issues (e.g., deficiencies and progress) on a regular basis?
- Is there a proper mechanism for new equipment to be reviewed by the plant safety engineer prior to its procurement or before it is put into operation?

3.8 Safety Cost Estimation

Over the years many methods and models have been developed to estimate various types of safety costs. This section presents one model and two such methods considered useful for application in the area of mine safety.

3.8.1 Safety Cost Estimation Model

This model is concerned with estimating total safety cost. The total safety cost is expressed by [12, 14]:

$$TSC = C_1 + C_2 + C_3 + C_4 + C_5 + C_6 + C_7 + C_8 \tag{3.1}$$

where

TSC is the total safety cost.
C_1 is the cost associated with accident prevention measures.
C_2 is the cost of immediate losses due to accidents.
C_3 is the cost associated with restoration and rehabilitation.
C_4 is the cost of insurance.
C_5 is the cost associated with safety-related legal issues.
C_6 is the cost associated with welfare issues.
C_7 is the cost associated with miscellaneous safety issues.
C_8 is the cost of immeasurables.

Additional information on this model is available in [12].

3.8.2 The Heinrich Method

This is a useful safety cost estimation method named after H.W. Heinrich, who pointed out that for each insured cost dollar paid for accidents, there were approximately four dollars of uninsured cost borne by the organization [15]. His conclusions were principally based on the following factors [16]:

- Evaluation of 5000 case files from companies or organizations insured with a private company
- Interviews with the administrative and production service personnel of these organizations
- Performance of various types of research studies in these enterprises

Heinrich divided the "total cost of occupational injuries" into two main elements: direct cost and indirect cost. The direct cost is made up of the total benefits paid by the insurance company, whereas the expense assumed directly by the organization is the indirect cost. More specifically, the direct cost is made up of elements such as listed below [15, 16]:

- Cost of injured employees' or workers' lost time
- Cost associated with material/machine damage
- Cost of management personnel's lost time
- Cost associated with profit and worker productivity loss
- Cost associated with lost orders
- Cost associated with overheads for injured workers while in non-production mode
- Cost associated with lost time of workers who stop their work and get involved in the action
- Cost associated with weakened morale
- Lost time cost of first aid and hospital workers not covered by insurance

Additional information on this method is available in [15, 16].

3.8.3 The Simonds Method

This method is named after R.H. Simonds, who developed it by reasoning that the total cost of an accident is made up of two main elements: insured cost and uninsured cost [17]. Furthermore, Simonds stressed that the insured cost can be estimated by simply examining a few accounting records or data, but the estimation of the uninsured cost is a challenging task. Thus, to calculate the uninsured cost of accidents, he developed the following equation [6, 17, 18]:

$$UC_a = \theta_1 C_1 + \theta_2 C_2 + \theta_3 C_3 + \theta_4 C_4 \quad (3.2)$$

where

UC_a is the uninsured cost of accidents.
θ_1 is the number of lost work-day cases due to class I accidents leading to temporary total disabilities and permanent partial disabilities.
C_1 is the average uninsured cost of Class I accidents.
θ_2 is the number of physician's cases associated with Class II accidents. More specifically, it is the OSHA non-lost workday cases requiring treatment by a doctor.
C_2 is the average uninsured cost of Class II accidents.
θ_3 is the number of first aid cases associated with Class III accidents. More specifically, those accidents in which first aid was given locally and which lead to a loss of less than one eighth of working time.
C_3 is the average uninsured cost of Class III accidents.
θ_4 is the number of non-injury cases associated with Class IV accidents. More specifically, those accidents that caused only minor injuries that did not require the attention of a healthcare professional.
C_4 is the average uninsured cost of Class IV accidents.

Additional information on this method is available in [6, 17].

3.9 Problems

1. Discuss at least seven principles of safety management.
2. What are the functions of a safety department?
3. Discuss a safety manager's functions.
4. Discuss the steps that can be used to develop a safety program plan for an organization.
5. Discuss the management-related deficiencies that can lead to accidents.
6. Discuss safety-related responsibilities of at least six non-safety groups in an organization.
7. List at least ten common questions asked in safety checklists used by management.
8. Describe the following two safety cost estimation methods:
   - The Heinrich method
   - The Simonds method
9. What are the main functions of a safety engineer?
10. Write an essay on safety management.

References

1. Petersen D. Safety Management: A Human Approach. Englewood, New Jersey: Aloray Publisher; 1975.
2. Petersen D. Safety Management. American Society of Safety Engineers. Des Plaines, Illinois; 1998.
3. Dhillon BS. Reliability Engineering in System Design and Operation. New York: Van Nostrand Reinhold Company; 1983.
4. Dhillon BS. Reliability and Quality Control: Bibliography on General and Specialized Areas. Gloucester, Ontario, Canada: Beta Publishers Inc.; 1992.
5. Grimaldi JV, Simonds RH. Safety Management. Chicago: Richard D. Irwin; 1989.
6. Goetsch DL. Occupational Safety and Health. Englewood Cliffs, New Jersey: Prentice Hall; 1996.
7. Roland HE, Moriarty B. System Safety Engineering and Management. New York: John Wiley and Sons; 1983.
8. Petersen D. Techniques of Safety Management. New York: McGraw Hill Book Company; 1971.
9. Gloss DS, Wardle MG. Introduction to Safety Engineering. New York: John Wiley and Sons; 1984.
10. Dhillon BS. Engineering Safety: Fundamentals, Techniques, and Applications. River Edge, New Jersey: World Scientific Publishing; 2003.
11. Cordtz D. Safety on the Job Becomes a Major Job for Management. Fortune 1972;11:112.
12. Hammer W, Price D. Occupational Safety Management and Engineering. Upper Saddle River, New Jersey: Prentice Hall, Inc.; 2001.
13. Peters T. Thriving on Chaos: Handbook for a Management Revolution. New York: Harper and Row; 1987.
14. Dhillon BS. Reliability, Quality, and Safety for Engineers. Boca Raton, Florida: CRC Press; 2005.
15. Heinrich HW. Industrial Accident Prevention. New York: McGraw Hill Book Company; 1931.
16. Andreoni D. The Cost of Occupational Accidents and Diseases. Geneva, Switzerland: International Labour Office; 1986.
17. Simonds R.M. Estimating Accident Cost in Industrial Plants. National Safety Council Safety Practices [pamphlet] No. III. Chicago; 1950.
18. Raouf A, Dhillon BS. Safety Assessment: A Quantitative Approach. Boca Raton, Florida: Lewis Publishers; 1994.

Chapter 4
Safety Analysis Methods and Indices

4.1 Introduction

The purpose of any safety analysis is typically to prevent the occurrence of potential accidents. This can only be satisfied effectively if the analysis is performed by using the most effective method, technique, or index for a stated problem. Over the years, many methods and indices have been developed to perform various types of analysis in areas such as reliability, safety, human factors, and quality control. Some of these methods are being used quite effectively across many diverse areas such as engineering design, management, healthcare, and maintenance.

Two examples of these methods are failure modes and effect analysis (FMEA) and fault tree analysis (FTA). FMEA was developed by the United States Department of Defense in the early 1950s to analyze engineering systems from the reliability aspect [1, 2]. It is currently being used in many diverse areas including safety, health care, and management to perform various types of analysis. FTA was developed in the early 1960s at the Bell Telephone Laboratories to analyze the Minuteman Launch Control System from reliability and safety perspectives [2]. Today, it is being used widely in industry to perform analysis of problems ranging from engineering-related to management-related.

This chapter presents a number of methods and indices extracted from published literature that are considered useful to perform various types of analysis in the area of mine safety.

4.2 Hazards and Operability (HAZOP)

This is a systematic approach used to identify hazards and operating-related problems throughout a facility under consideration. Past experiences indicate that

HAZOP is an extremely effective tool to highlight unforeseen hazards designed into facilities due to various reasons, or introduced into existing facilities due to factors such as changes made to process conditions or operational procedures.

Three fundamental objectives of this method are as follows [3]:

- **Objective I:** to produce a complete facility/process description
- **Objective II:** to review every facility/process element to determine how deviations from the design intentions can occur
- **Objective III:** to decide whether such deviations can lead to operating hazards/problems

A HAZOP study is conducted by following the five steps listed below [3–5]:

- **Step 1.** Establish study objectives and scope
  This step is concerned with developing the objectives and scope of the study by considering relevant factors
- **Step 2.** Form HAZOP team
  This step is concerned with forming a HAZOP team by insuring that all team members are comprised of individuals from design and operation with expertise to determine all types of deviation effects from intended applications.
- **Step 3.** Collect relevant information
  This step is concerned with obtaining the necessary process descriptions, drawings, and documentation. More specifically, this includes items such as process flow-sheets, process control logic diagrams, equipment specifications, layout drawings, operating and maintenance procedures, and emergency response procedures.
- **Step 4.** Conduct analysis of all major pieces of equipment and supporting items
  This step is concerned with performing analysis of all major items of equipment, and all supporting equipment, piping, and instrumentation by using the documents of the preceding steps.
- **Step 5.** Document the study
  This step is concerned with documenting items such as the consequences of any deviation from the norm, a summary of any deviation from the norm, and a summary of deviations considered credible and hazardous.

Additional information on this method is available in [3–5].

4.3 Job Safety Analysis (JSA)

This is one of the safety management tools used for uncovering and rectifying potential hazards which are intrinsic to or inherent in the workplace. Normally worker, safety professional, supervisor, and management are involved in performing JSA and the analysis is conducted by following the five steps listed below [5–7]:

- **Step 1.** Select a job for analysis
- **Step 2.** Split the job under consideration into a number of tasks or steps
- **Step 3.** Identify all potential hazards and determine suitable measures to control these hazards
- **Step 4.** Apply Step 3 measures to control hazards
- **Step 5.** Evaluate all controls with care

Finally, it is emphasized that the degree of success of this method greatly depends upon the rigor that the JSA team members exercise during the analysis process. Additional information on JSA is available in [5–7].

4.4 Preliminary Hazard Analysis (PHA)

PHA is generally utilized during the conceptual design phase and is a relatively unstructured method because of the unavailability of definitive information such as functional flow diagrams and drawings. This method has proven to be an effective approach to take early steps to identify and eliminate potential hazards when the required data are not available. Thus, the results of PHA are very useful to serve as a guide in potential detailed analysis.

The method involves the formation of an *ad hoc* team composed of individuals who are familiar with items such as equipment, substances, materials, and/or processes under consideration. All members of this team are asked to review the occurrence of hazards in the area of their expertise, and as a team these members play the devil's advocate.

The findings of this analysis are used for purposes such as preparing specification, testing, implementation, and maintenance. Additional information on PHA is available in [5, 8].

4.5 Failure Mode and Effective Analysis

This is a widely used method in the industrial sector to analyze the reliability aspects of engineering systems. More specifically, FMEA is used to analyze each potential failure mode in a given system to determine the effects of such failure mode on the system [9].

The history of this method may be traced back to the early years of the 1950s when the Bureau of Aeronautics of the United States Navy developed a requirement called "Failure Analysis" in order to establish a procedure for reliability control over the detail design effort [10]. In later years, the term "Failure Analysis" was changed to failure mode and effect analysis (FMEA) and FMEA is called failure modes, effects, and criticality analysis (FMECA) when criticalities or priorities are assigned to failure mode effects [11].

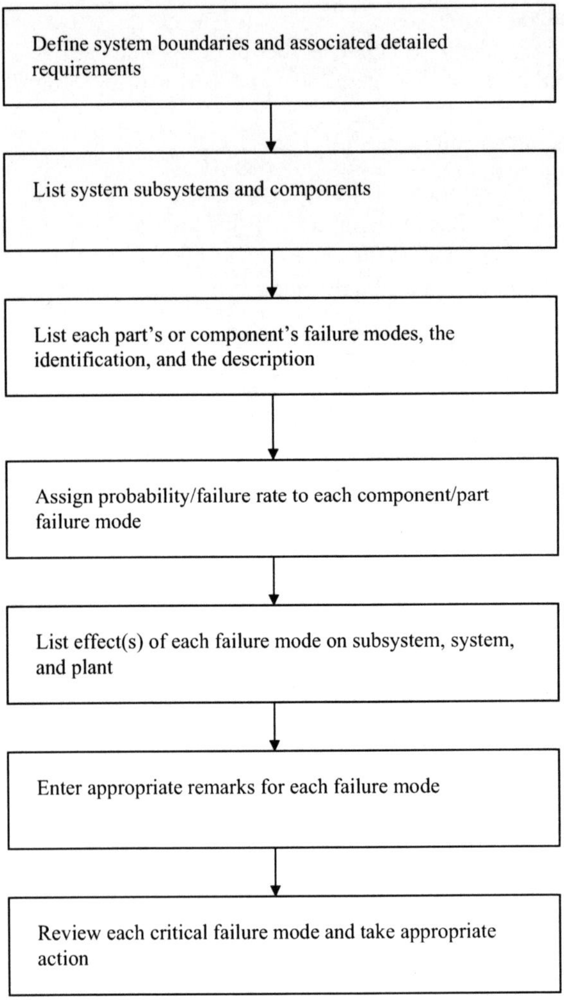

Figure 4.1 FMEA main steps

The main steps followed to perform FMEA are shown in Figure 4.1 [2]. Some of the benefits of performing FMEA are as follows [12]:

- An effective tool for comparing designs
- Highlights safety-related concerns to be focused on
- Easy to understand and improves customer satisfaction
- A useful approach that starts from the detailed level and works upward
- Reduces development time and cost
- A visibility tool for management

- An effective approach to improve communication among design interface personnel
- A systematic method to classify hardware failures

Additional information on this method is available in [2, 12].

4.6 Interface Safety Analysis (ISA)

This method is concerned with determining the incompatibilities between assemblies and subsystems of a system that could result in accidents. The method establishes that totally distinct elements or units can be integrated into a viable system and the normal functioning of an individual unit will not impair the performance or damage another unit or the system.

ISA considers a wide variety of relationships that nonetheless can be classified principally into three areas as shown in Figure 4.2 [13].

The functional relationships are concerned with multiple items. For example, in a situation where outputs of a unit constitute the inputs to a downstream item or unit, any error in outputs and inputs may cause damage to the downstream item or unit, and consequently a safety hazard. The outputs could be in conditions such as degraded outputs, zero outputs, excessive outputs, and erratic outputs.

Flow relationships are concerned with two or more items or units. For example, the flow between two items or units may involve air, steam, water, fuel, lubricating oil, or electrical energy, and it could be unconfined as in the radiation of heat from one body to another. Past experience indicates that frequent problems faced with many products involve the proper flow of fluids and energy from one item/unit to another through confined spaces or passages that consequently result in safety-related problems. Flow-related problem causes include faulty connections between units and total or partial interconnection failure.

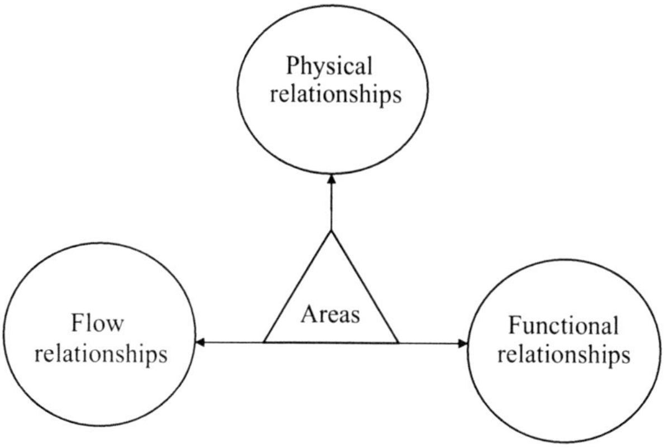

Figure 4.2 Areas of relationships considered by ISA

Physical relationships are concerned with the physical aspects of items or units. For example, two units or items might be well designed and manufactured and operate quite well individually, but they may not fit together because of dimensional differences or they may cause other problems that could lead to safety-related difficulties. Three specific examples of other problems are as follows [5, 13]:

- A very small clearance between units or items, possibly resulting in damage to the units or items during the removal process
- Impossible or restricted access to or egress from equipment or system
- Impossibility to mate, tighten, or join parts properly

Additional information on ISA is available in [13].

4.7 Fault Tree Analysis

This is a widely used method in the industrial sector to perform reliability analyses of engineering systems during their design and development, particularly in the area of nuclear power generation. The method was developed in the early 1960s to perform reliability analyses of the Minuteman Launch Control System [2, 14]. Some of the common objectives of performing fault tree analysis are as follows [2]:

- To confirm system reliability
- To confirm the ability of the system to satisfy its specified safety requirements
- To identify critical areas and cost-effective improvements
- To understand the functional relationship of system failures

FTA starts by identifying an undesirable event, known as a top event, associated with a system under study. Fault events that can cause the occurrence of the top event are generated and connected by logic operators such as AND and OR. The AND gate provides a True output (i.e., fault) if all its inputs are True (fault) and the OR gate provides a True output (fault) if one or more of its inputs are True (fault).

The construction of a fault tree proceeds by generating fault events successively until the fault events need not be developed any further. These fault events are called basic or primary events. A fault tree is a logic structure that relates the top fault event to the primary fault events. During the construction process of a fault tree, one question successively asked is, "How could this fault event occur?"

Four commonly used symbols in the construction of fault trees are shown in Figure 4.3 [2, 14].

The meanings of symbols/gates AND and OR were already described above. The rectangle represents a resultant fault event that occurs from the combination of fault events through the inputs of a logic gate such as AND or OR. The circle denotes a basic or primary fault event (e.g., failure of an elementary part), and the parameters of a basic fault event are failure probability, failure and repair rates, and unavailability.

The construction of a fault tree is demonstrated through the following example.

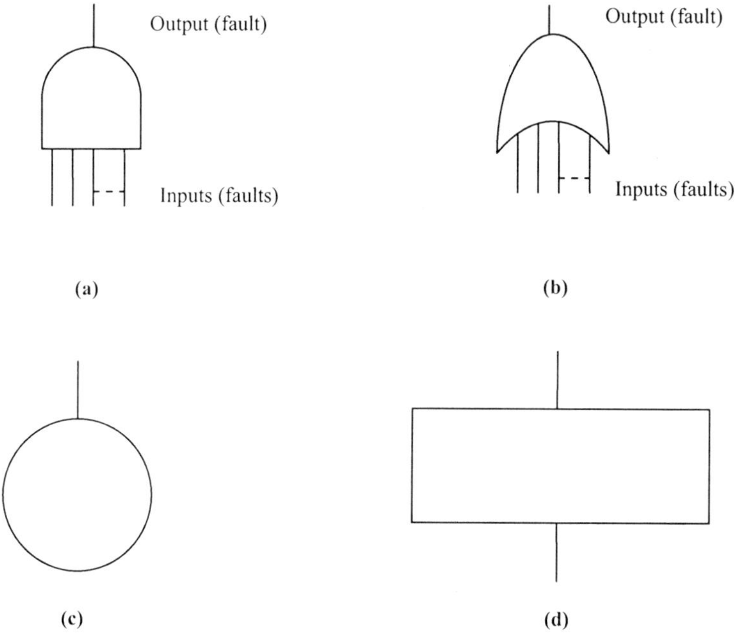

Figure 4.3 Common fault tree symbols: (**a**) AND gate, (**b**) OR gate, (**c**) basic fault event, and (**d**) resultant event

Example 4.1

A windowless room contains two light bulbs (*X* and *Y*) and one switch that can only fail to close. Develop a fault tree for the occurrence of the top event "dark room" by using Figure 4.3 symbols.

In this case, the room can only be dark when there is no incoming electricity, the switch fails to close, or both the light bulbs (*X* and *Y*) burn out. A fault tree for the example is shown in Figure 4.4.

4.7.1 Probability Evaluation of Fault Trees

It occasionally becomes necessary to predict the probability of occurrence of certain undesirable or unsafe top events. Before this can be achieved by using the FTA method, it is absolutely essential that an evaluation of the probability of occurrence of output fault events of the logic gates is performed. Thus, the probability of occurrence of the output fault event of an OR gate is expressed by [2]:

$$P_{org}(E) = 1 - \prod_{j=1}^{m} (1 - P(E_j)) \quad (4.1)$$

where

$P_{org}(E)$ is the probability of occurrence of the OR gate's output fault event, E.

$P(E_j)$ is the probability of occurrence of input fault event E_j, for $j = 1, 2, 3, ---, m$.

m is the number of input fault events.

Similarly, the probability of occurrence of the output fault event of an AND gate is expressed by

$$P_{oag}(E) = \prod_{j=1}^{m} P(E_j) \tag{4.2}$$

where

$P_{oag}(E)$ is the probability of occurrence of the AND gate's output fault event E.

Example 4.2

Assume that in Example 4.1 the probabilities of occurrence of basic fault events "power failure", "fuse failure", "bulb *X* burnt out", "bulb *Y* burnt out", and "switch fails to close" are 0.01, 0.04, 0.07, 0.08, and 0.02, respectively. Calculate the probability of occurrence of the top event "dark room" by using Equations 4.1 and 4.2.

After substituting the given data values into Equation 4.1, the occurrence probability of the event "No electricity" is

$$\begin{aligned} P_{ne} &= 0.01 + 0.04 - (0.01)(0.04) \\ &= 0.0496 \end{aligned}$$

where

P_{ne} is the probability of occurrence of the event "No electricity".

Using the specified data values in Equation 4.2, we obtain the following value for the probability of occurrence of the event "Both bulbs burnt out":

$$P_{bb} = (0.07)(0.08) = 0.0056$$

By inserting the calculated values and the data value given above into Equation 4.1, we get the following value for the probability of occurrence of the top event "Dark room":

$$\begin{aligned} P_{dr} &= 1 - (1 - 0.0056)(1 - 0.0496)(1 - 0.02) \\ &= 0.0738 \end{aligned}$$

Thus, the probability of occurrence of the top event "Dark room" is 0.0738.

The fault tree of Figure 4.4 with the calculated probability values and given event occurrence probability values is shown in Figure 4.5.

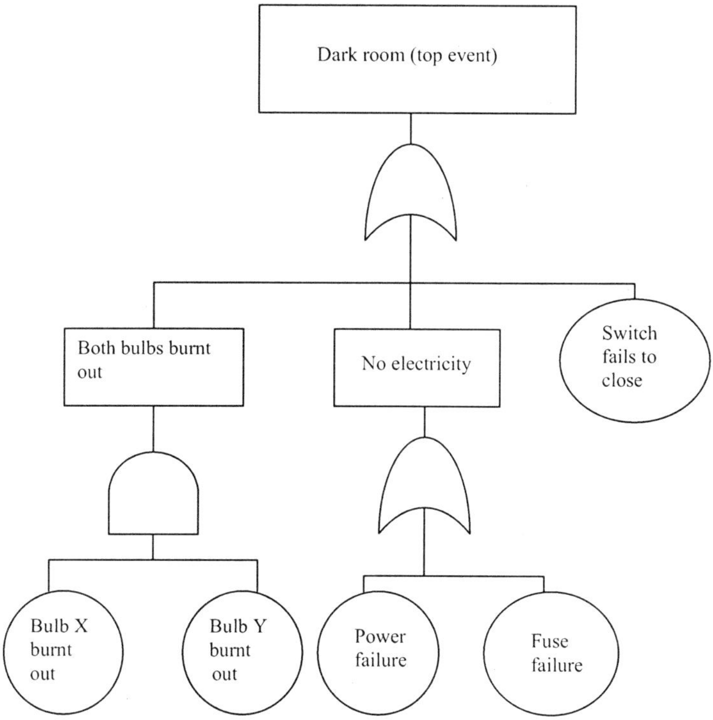

Figure 4.4 Fault tree for Example 4.1

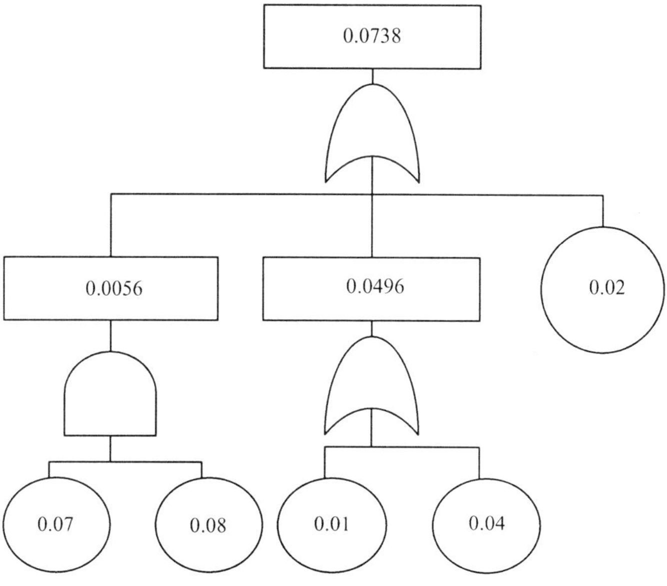

Figure 4.5 Redrawn Figure 4.4 fault tree with given and calculated event occurrence probability values

4.7.2 Fault Tree Analysis Benefits and Drawbacks

There are many benefits and drawbacks of the FTA method. Some of its benefits are as follows [2, 14]:

- Useful to handle complex systems more easily
- Allows concentration on one fault or failure at a time
- Useful in providing insight into the system behavior
- Useful for providing options to management and others to carry out either quantitative or qualitative analysis
- Forces the involved analyst to understand the system thoroughly prior to starting the analysis
- Useful as a graphic aid for management and others and in the deductive identification of failures/faults

In contrast, some of the drawbacks of FTA are that it is a costly and time consuming approach, its end results are difficult to check, and it considers components in either working or failed states [2, 14]. Additional information on FTA is available in [2, 14].

4.8 Markov Method

This method is named after a Russian mathematician, Andrei A. Markov (1856–1922), and is often used to perform reliability analyses of engineering systems. The method can also be used to perform safety analyses in the area of mining. The following assumptions are associated with the Markov method [2, 15]:

- All occurrences are independent of each other.
- The probability of transition from one system state to another in the finite time interval Δt is given by $\theta\Delta t$, where the θ is the transition rate (e.g., failure or repair rate) from one system state to another.
- The probability of more than one transition occurrence in the finite time interval Δt from one system state to another is very small or negligible (i.e., $(\theta\Delta t)(\theta\Delta t) \rightarrow 0$).

The application of this method to a mine safety-related problem is demonstrated through the following example.

Example 4.3

A mining system can fail either safely or unsafely and its safe and unsafe failure rates are λ_1 and λ_2, respectively. The state space diagram of the mining system is shown in Figure 4.6. The numerals in boxes denote the mining system states.

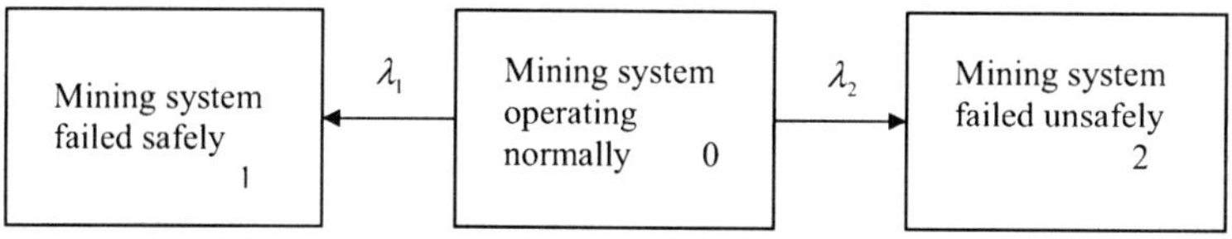

Figure 4.6 Mining system state space diagram

Develop expressions for the mining system state probabilities and mean time to failure by using the Markov method. Assume that the mining system is subjected to the assumptions that system failures occur independently and the system safe and unsafe failure rates are constant.

By using the Markov method, we write down the following equations for states 0, 1, and 2, respectively, shown in Figure 4.6:

$$P_0(t+\Delta t) = P_0(t)(1-\lambda_1\Delta t)(1-\lambda_2\Delta t) \tag{4.3}$$

$$P_1(t+\Delta t) = P_1(t)(1-0\Delta t) + P_0(t)\lambda_1\Delta t \tag{4.4}$$

$$P_2(t+\Delta t) = P_2(t)(1-0\Delta t) + P_0(t)\lambda_2\Delta t \tag{4.5}$$

where

- t is time.
- $P_0(t+\Delta t)$ is the probability of the mining system being in operating state 0 at time $(t+\Delta t)$.
- $P_1(t+\Delta t)$ is the probability of the mining system being in safe failed state 1 at time $(t+\Delta t)$.
- $P_2(t+\Delta t)$ is the probability of the mining system being in unsafe failed state 2 at time $(t+\Delta t)$.
- $P_j(t)$ is the probability that the mining system is in state j at time t, for $j=0$ (operating normally), $j=1$ (failed safely), and $j=2$ (failed unsafely).
- $\lambda_1\Delta t$ is the probability of safe mining system failure in finite time interval Δt.
- $\lambda_2\Delta t$ is the probability of unsafe mining system failure in finite time interval Δt.
- $(1-\lambda_1\Delta t)$ is the probability of no safe mining system failure in finite time interval Δt.
- $(1-\lambda_2\Delta t)$ is the probability of no unsafe mining system failure in finite time interval Δt.

Using Equation 4.3, we get

$$P_0(t+\Delta t) = P_0(t)\left[1-\lambda_1\Delta t-\lambda_2\Delta t+(\lambda_1\Delta t)(\lambda_2\Delta t)\right] \tag{4.6}$$

Since $(\lambda_1\Delta t)(\lambda_2\Delta t)\rightarrow 0$, Equation 4.6 reduces to

$$P_0(t+\Delta t) = P_0(t)\left[1-\lambda_1\Delta t-\lambda_2\Delta t\right] \tag{4.7}$$

Using Equation 4.7, we write

$$\lim_{\Delta t \to 0} \frac{P_0(t+\Delta t) - P_0(t)}{\Delta t} = -\lambda_1 P_0(t) - \lambda_2 P_0(t) \tag{4.8}$$

Thus, from Equation 4.8, we get

$$\frac{dP_0(t)}{dt} + \lambda_1 P_0(t) + \lambda_2 P_0(t) = 0 \tag{4.9}$$

Similarly, using Equations 4.4 and 4.5, we get the following equations:

$$\frac{dP_1(t)}{dt} - \lambda_1 P_0(t) = 0 \tag{4.10}$$

$$\frac{dP_2(t)}{dt} - \lambda_2 P_0(t) = 0 \tag{4.11}$$

At time $t = 0$, $P_0(0) = 1$, $P_1(0) = 0$, and $P_2(0) = 0$.
By solving Equations 4.9–4.11, we obtain

$$P_0(t) = e^{-(\lambda_1+\lambda_2)t} \tag{4.12}$$

$$P_1(t) = \frac{\lambda_1}{\lambda_1+\lambda_2}\left[1 - e^{-(\lambda_1+\lambda_2)t}\right] \tag{4.13}$$

$$P_2(t) = \frac{\lambda_2}{\lambda_1+\lambda_2}\left[1 - e^{-(\lambda_1+\lambda_2)t}\right] \tag{4.14}$$

By integrating Equation 4.12 over the time interval [0, ∞], we obtain the following equation for the mining system mean time to failure [2]:

$$MTTF_{ms} = \int_0^\infty e^{-(\lambda_1+\lambda_2)t}\, dt \tag{4.15}$$

where $MTTF_{ms}$ is the mining system mean time to failure.

Example 4.4

Assume that a mining system's safe and unsafe failure rates are 0.009 failures/h and 0.001 failures/h, respectively. Calculate the mining system probability of failure due to an unsafe failure during a 20-h mission and the mean time to failure.

By substituting the given data values into Equations 4.14 and 4.15, we get

$$P_2(20) = \frac{0.001}{0.009+0.001}\left[1 - e^{-(0.009+0.001)(20)}\right]$$
$$= 0.0181$$

and

$$MTTF_{ms} = \frac{1}{0.009 + 0.001}$$
$$= 100\,\text{h}$$

Thus, the mining system's probability of failure due to an unsafe failure and the mean time to failure are 0.0181 and 100 hr, respectively.

4.9 Index I: Disabling Injury Severity Rate (DISR)

This index was proposed by the American National Standards Institute (ANSI) and is based on four factors occurring during the period covered by the rate (i.e., total scheduled charges (days) for all deaths, permanent total and permanent partial disabilities, and the number of days of disability from all temporary injuries) [16].

The index is expressed by

$$DISR = \frac{DC(1,000,000)}{T_{ee}} \tag{4.16}$$

where

DC is the number of days charged.
T_{ee} is the employee exposure time in hours.

Three important advantages of this index are as follows [16]:

- As an effective tool for answering the question, "How serious are injuries in our organization?"
- As an effective tool for taking into consideration differences in quantity of exposure over time.
- As an effective tool for making a meaningful comparison between different organizations.

4.10 Index II: Disabling Injury Frequency Rate (DIFR)

This index was also proposed by the American National Standards Institute and is based on four factors/events that occur during the time period covered by the rate (i.e., temporary disabilities, permanent partial disabilities, permanent disabilities, and deaths).

The index is expressed by

$$DIFR = \frac{\alpha(1,000,000)}{T_{ee}} \tag{4.17}$$

where

α is the number of disabling injuries.

One main advantage of DIFR is that it clearly considers differences in quantity of exposures due to varying employee work hours, either among organizations categorized under the similar industry group or within the organization during successive periods [17, 18].

4.11 Problems

1 What are the basic objectives of hazards and operability analysis?
2 Describe job safety analysis.
3. Discuss preliminary hazard analysis.
4. What are the advantages of performing failure mode and effect analysis?
5. What is interface safety analysis?
6. What are the main advantages of the fault tree analysis method?
7. What are the four commonly used symbols in the construction of fault trees? Describe each of these symbols.
8. Prove that the sum of Equations 4.12–4.14 is equal to unity.
9. Prove Equation 4.14 by using Equations 4.9–4.11.
10. Define the following two indices:

    - DISR
    - DIFR

References

1. Continho JS. Failure Effect Analysis. Trans NY Acad Sci 1963–1964; 26:564–584.
2. Dhillon BS. Design Reliability: Fundamentals and Applications. Boca Raton, Florida: CRC Press; 1999.
3. Canadian Standards Association. Risk Analysis Requirements and Guidelines. CAN/CSA-Q6340-91; 1991. Available from Canadian Standards Association (CSA), 178 Rexdale Boulevard, Rexdale, Ontario, Canada.
4. Gould J, Glossop M, Ioannides A. Review of Hazard Identification Techniques. Report No. HSL/2005/58. London: Health and Safety Executive; 2000.
5. Dhillon B.S. Engineering Safety: Fundamentals, Techniques, and Applications. New York: World Scientific Publishing; 2003.
6. Hammer W, Price D. Occupational Safety Management and Engineering. Upper Saddle River, New Jersey: Prentice Hall, Inc.; 2001.
7. Kjellen U. Prevention of Accidents through Experience Feedback. Boca Raton, Florida: CRC Press; 2001.
8. Ericson II CA. Hazard Analysis Technique for System Safety. New York: John Wiley and Sons; 2005.
9. Omdahl TP, editor. Reliability, Availability, and Maintainability (RAM) Dictionary. Milwaukee, Wisconsin: American Society for Quality Control (ASQC) Press; 1988.
10. MIL-F-18372 (Aer), General Specification for Design, Installation, and Test of Aircraft Flight Control Systems. Bureau of Naval Weapons Department of the Navy. Washington, D.C.

11. MIL-STD-1629, Procedures for Performing a Failure Mode, Effects, and Criticality Analysis. Department of Defense, Washington, D.C.; 1980.
12. Palady P. Failure Modes and Effects Analysis. West Palm Beach, Florida: PT Publications; 1995.
13. Hammer W. Product Safety Management and Engineering. Englewood Cliffs, New Jersey: Prentice Hall, Inc.; 1980.
14. Dhillon BS, Singh C. Engineering Reliability: New Techniques and Applications. New York: John Wiley and Sons; 1981.
15. Shooman ML. Probabilistic Reliability: An Engineering Approach. New York: McGraw-Hill Book Company; 1968.
16. Z16.1, Method of Recording and Measuring Work Injury Experience. New York: American National Standards Institute; 1985.
17. Tarants WE. The Measurement of Safety Performance. New York: Garland STPM Press; 1980.
18. Dhillon BS. Reliability, Quality, and Safety for Engineers. Boca Raton, Florida: CRC Press; 2005.

Chapter 5
Global Mine Accidents

5.1 Introduction

A mining accident may simply be defined as an accident that occurs in the process of mining minerals from underneath the surface of the planet. Each year, thousands of miners die from mining accidents, particularly in the area of coal mining and hard rock mining.

There are various causes for the occurrence of mining accidents, including leaks of poisonous or explosive natural gases, collapsing of mine stopes, dust explosions, flooding, or general mechanical errors from incorrectly used or malfunctioning mining equipment [1, 2].

Each year, a large number of fatalities occur in mines globally. Most of these fatalities occur in developing countries and rural parts of developed countries. Nonetheless, even in the United States an average of 93 people died in mining accidents during the period 1991–1999, in addition to an average of 21, 351 injuries per year [3]. Currently, China accounts for a large proportion of mining accident related fatalities, particularly in the area of coal mining. For example, it produces around 35% of the world's coal and accounts for about 80% of coal-mining fatalities.

Also, the worst coal mining disaster in the world occurred in China on April 26, 1942 at the Benxihu Colliery, located at Benxi, Liaoning. In this mining accident, a coal-dust explosion killed 1572 people [4]. This chapter presents various important aspects of global mine accidents.

5.2 Mine Accidents: United States

Mining is one of the largest industries in the United States in terms of the number of ongoing operations, as well as by volume and weight. For example, in 2003 the value of non-fuel mineral commodities produced in the United States was estimated to be approximately $38 billion [5]. The history of documented mine accidents or disasters in the United States may be traced back to the nineteenth century. In fact, during the period 1876–1900 there were 118 mine disasters in coal and metal/non-metal mines that claimed 5 or more deaths [6].

In 1907, the deadliest mining disaster in the United States history occurred in a coal mine near Monongah, West Virginia, in which 362 persons were killed [6]. During the period 1908–1935, many other deadly mining accidents occurred in the

Table 5.1 Breakdown of average number of mining deaths per year in 5-year periods for the period 1935–2005

Period No.	Time period	Average number of deaths per year
1	1936–1940	1546
2	1941–1945	1592
3	1946–1950	1054
4	1951–1955	690
5	1956–1960	550
6	1961–1965	449
7	1966–1970	426
8	1971–1975	322
9	1976–1980	254
10	1981–1985	174
11	1986–1990	122
12	1991–1995	99
13	1996–2000	86
14	2001–2005	62

Table 5.2 Yearly breakdowns of mining fatalities in the United Sates for the period 1996–2008

Year	No. of mining fatalities
1996	85
1997	91
1998	80
1999	90
2000	85
2001	72
2002	69
2003	56
2004	55
2005	58
2006	73
2007	67
2008	52

United States including the Cherry, Illinois and Mather, Pennsylvania mine disasters in 1909 and 1928, respectively. In these disasters 259 and 195 persons were killed, respectively.

During the period 1936–2005, a large number of deaths in the United States mining industry have occurred. Their 5 year periodic breakdowns, i.e., the average number of deaths per year during a given 5-year period are presented in Table 5.1 [7].

Mining fatalities that occurred each year in the United States during the period 1996–2008 are presented in Table 5.2 [8].

It is to be noted from Tables 5.1 and 5.2 that in the early part of the 20th century there were over 1000 mining fatalities per year, and they reduced to the average of about 65 per year for the period 2000–2008.

It should also be noted that over the years the fatality rate in the United States in coal, metal, and non-metal mines has considerably decreased. In coal mines, it decreased from around 0.20 fatalities per 200,000 hr worked by miners (or one death per million production hours) in 1970 to around 0.07 fatalities in 1977, and to an average of about 0.03 fatalities for the period 2001–2005. Similarly, the metal and non-metal mining fatality rate per 200,000 employee hours averaged around 0.02 for the period 2001–2005, compared to annual rates about three times higher in the 1950s and seven times as high in the 1930s [7].

There has been a combination of factors during this time period that is responsible for the dramatic safety-related gains in the United States mining industry. The main elements of these gains have been the following [7]:

- The creation of the U.S. Bureau of Mines by the United States Congress in 1910. The primary role of this body was to advise industry, investigate accidents, conduct production and safety-related research, and teach appropriate courses in areas such as accident prevention, mine rescue, and first aid.
- Passage of federal and state laws to better advise and regulate the mining industrial sectors to encourage/require application of effective safety- related technology and procedures, to focus on eliminating or reducing the most serious hazards, to extend coverage to all types of miners, and to provide appropriate and effective miner training. The federal Coal Mine Health and Safety Act of 1969 and the Federal Mine Safety and Health Act of 1977 were the most far-reaching laws.
- Creation of the Mining Enforcement and Safety Administration in 1973 within the framework of the Interior Department. This body assumed safety and health enforcement responsibilities from the Bureau of Mines.
- Creation of the current Mine Safety and Health Administration (MSHA) in 1977 and its transfer to the Labor Department.
- Introduction of safer mining systems and methods, a growing awareness of the importance of appropriate and effective accident prevention programs by both miners and management personnel, and a more cooperative attitude toward safety-related issues by government, labor, and the mining industrial sector.

Two of the top ten deadliest mining disasters in the United States, are described below, separately.

5.2.1 Monongah Mining Disaster

This disaster occurred at Monongah, West Virginia on December 6, 1907 and has been described as "the worst mining disaster in the history of the United States" because it caused 362 fatalities. The disaster was the result of an explosion in mines 6 and 8 of the Fairmont Coal Company of Fairmont, West Virginia. It is generally believed that the explosion was caused by the ignition of methane, which in turn ignited the coal dust in these two mines [9, 10].

The United States Congress reacted to this disaster by passing and toughening mining laws. Additional information on the Monongah mining disaster is available in [9–12].

5.2.2 Cherry Mine Disaster

This disaster occurred at Cherry, Illinois on November 13, 1909 and is considered one of the top ten mining disasters in the United States, causing 259 fatalities. The Cherry Mine was opened in 1905 to supply coal for trains and it was composed of three horizontal veins. Two shafts that were set about 300 ft apart vertically connected the veins. Both the shafts contained wooden stairs and ladders.

A coal car filled with hay for the mules accidentally caught fire from one of the wall lanterns, and it in turn set the entire coal mine on fire [13].

In 1910, as a result of this disaster, the Illinois legislature established stronger fire and safety-related regulations governing mines in the state. In 1911, Illinois passed a liability act, which subsequently developed into the Illinois Workmen's Compensation Act. Additional information on Cherry Mine Disaster is available in [13, 14].

5.3 Mine Accidents: United Kingdom

Mining has been and still is, to a certain degree, an important element of the industrial sector in the United Kingdom. There are around 137 mining sites still in operation, including 26 licensed underground coal mines and 31 opencast coal sites. The death toll from mining accidents in the country has fallen from one in every 1000 employees at the beginning of the last century to an average of less than one for the period 2003–2007.

Some of the largest and worst mining accidents in the United Kingdom occurred in Wales. Over the period 1850 to 1930 the South Wales coalfield had the worst disaster record, mainly due to factors such as increasing number of mines sunk to greater depths into gas-containing strata and poor safety and management practices. The four worst mining accidents that occurred in Wales are as follows [15]:

- 259 fatalities at the Prince of Wales Mine, Abercarn in an explosion in 1878
- 266 fatalities in the Gresford Mine Disaster near Wrexham, North Wales in 1934
- 290 fatalities at the Albion Colliery in Cilfynydd, South Wales in a gas explosion in 1894
- 439 fatalities at the Universal Colliery in Senghenydd, South Wales in an explosion in 1913

The worst mining accident that occurred in Scotland was in 1877 known as the Blantyre mining disaster in Blantyre, Lanarkshire. It caused 207 fatalities. In England, the oaks explosion near Barnsley, Yorkshire in 1866 remains the worst mining accident in that country, claiming 388 lives. The two deadliest United Kingdom mining disasters are described below, separately.

5.3.1 Senghenydd Colliery Disaster

This mining disaster occurred in Senghenydd, South Wales on October 14, 1913 and is considered the worst mining accident in the United Kingdom because it caused 439 fatalities. This disaster is believed to be the result of a firedamp (methane) explosion. The methane probably ignited by an electric spark from equipment such as electric bell signaling gear. Additional information on this disaster is available in [16, 17].

5.3.2 The Oaks Mining Disaster

This was one of the worst mining accidents and colliery disasters in the United Kingdom and it occurred when the oaks pit exploded on December 12, 1866. The disaster killed 388 miners and their would-be rescuers. The Oaks was one of the largest coal mines in the rich Barnsley coal seam in Yorkshire, England. Its coal was considered to be very gassy.

Although various theories have been put forward for the cause of this disaster, the exact cause is still unclear. Additional information, on the Oaks mining disaster is available in [18].

5.4 Mine Accidents: China

Mining is an important element of the Chinese industrial sector. In fact, Chinese mines produce around 35% of the world's coal and in 2006 the country generated about 69% of its electricity using coal [15, 19]. Currently, about 80% of the world's coal mining fatalities occur in China.

In 1942, the deadliest mining disaster in the world occurred in Benxihu (Honkeiko) Colliery, located at Benxi, Liaoning province, in which 1549 miners were killed.

During the period January 2001–October 2004, 188 mining accidents with a death toll of more than 10 have occurred in the entire country. According to State Work Safety Supervision Administration, in 2006 4749 Chinese coal miners were killed in a large number of accidents [19, 20]. Two of the Chinese mining disasters are described below, separately.

5.4.1 Benxihu (Honkeiko) Colliery Mining Disaster

This colliery was located near Benxi Lake in the Ore-rich region of eastern Chinese province called Lianoning. The Colliery came into operation in 1905 and on April 26, 1942 an accident in the colliery killed 1549 workers, a total of 34% of the miners working that day.

A subsequent investigation into the accident pointed out that the disaster was caused by a gas and coal-dust explosion. This disaster is considered as the worst mining disaster in the world in the term of fatalities. Additional information on the disaster is available in [15, 19].

5.4.2 Sunjiawan Mine Disaster

Sunjiawan mine is located at Fuxin city, Liaonning province, People's Republic of China. On February 14, 2005, an accident in this mine resulted in at least 214 fatalities [20].

A subsequent investigation into the disaster revealed that the disaster was the result of a gas explosion that occurred about 10 min after an earthquake shook the mine. Additional information on the Sunjiawan mine disaster is available in [20].

5.5 Mine Accidents: Australia

Many different ores and minerals are mined in Australia including iron ore, nickel, copper, gold, silver, diamonds, coal, bauxite, zinc, uranium, and petroleum. Mining is a significant industry in Australia and it contributes around 6% of Australia's Gross Domestic Product.

Over the years many mining accidents have occurred in Australia that resulted in a number of deaths. Some of these were the Mount Kembla mining disaster, the Mount Mulligan mining disaster, the Bulli colliery disaster of 1887, the Balmain colliery disasters of 1900, 1932, and 1945, the North Mount Lyell mining disaster, and the Moura mining disasters of 1975, 1986, and 1994 [21]. Three of these mining disasters are described below, separately.

5.5.1 Mount Kembla Mining Disaster

The Mount Kembla coal mine is located in the Illawarra District of New South Wales, Australia and it was opened in 1883. On July 31, 1902, an accident in this mine resulted in 96 fatalities. In 1903, a Royal Commission investigating the disaster concluded that the disaster was caused by a gas and coal dust explosion.

This is the worst mining disaster in the history of Australia. Additional information on Mount Kembla mining disaster is available in [21, 22].

5.5.2 Mount Mulligan Mining Disaster

This mine is located at Mount Mulligan, Far North Queensland, Australia. The mining disaster occurred on September 19, 1921 and is considered as the third worst mining accident in Australia because it caused 75 fatalities.

A subsequent investigation by the Royal Commission into the disaster stated that the accident was caused by the detonation of firedamp. Furthermore, the commission pointed out that the explosives were carelessly carried underground, stored, distributed, and used. Additional information on the disaster is available in [21, 23].

5.5.3 Bulli Collliery Disaster

This colliery is located at Bulli, New South Wales, Australia. The colliery was a “gassy” pit with high concentrations of firedamp within its coal. On March 23, 1887, an accident in this colliery caused 81 deaths. A subsequent investigation into the accident pointed out that the disaster occurred because of an explosion caused by marsh gas or carbonic hydrate that had accumulated at the face. Additional information on the Bulli Colliery disaster is available in [21, 23].

5.6 Mine Accidents: Canada

Canada is one of the leading mining nations in the world. Its mining industry (including smelting and refining) contributes around 4% to nation’s Gross Domestic Product. The country is a major producer and exporter of various important commodities such as copper, nickel, gold, zinc, iron ore, and lead.

Over the years, many mining accidents have resulted in fatalities in Canada. Some of these were the Drummond Colliery mining disaster, the Westray mining disaster, the Springhill mining disaster, the Fernie mining disaster, the Hillcrest mining disaster, and the Nanaimo mining disaster that occurred in 1873, 1992, 1958, 1902, 1914, and 1887, respectively. Four of these mining disasters are described below, separately.

5.6.1 Hillcrest Mining Disaster

The Hillcrest coal mine is located at Hillcrest, Alberta, in the Crowsnest Pass region of Western Canada. The mining disaster occurred on June 19, 1914 and is considered the worst mining accident in Canada because it caused 189 fatalities. The disaster was caused by a large explosion in the mine. Additional information on the disaster is available in [24, 25].

5.6.2 Springhill Mining Disaster

The Springhill coal mine is in close proximity to the town of Springhill in Cumberland County, Nova Scotia, Canada. The mining disaster occurred on October 23, 1958 and it caused 74 fatalities. The disaster was the result of a severe "bump" (underground earthquake). Additional information on the disaster is available in [26].

5.6.3 Nanaimo Mining Disaster

Nanaimo mine is located at Nanaimo, British Columbia, Canada. On May 23, 1887 an accident in this mine resulted in 150 fatalities. The accident was caused by an explosion that started deep underground in the coal mine because explosives were laid improperly. Additional information on this disaster is available in [27].

5.6.4 Westray Mining Disaster

Westray is a coal mine located at Plymouth, Nova Scotia, Canada. The mine was opened on September 11, 1991.

The mining disaster occurred on May 9, 1992 and it caused 26 fatalities. The disaster was caused by methane gas and a subsequent coal dust explosion, and it was Canada's worst mining disaster since 1958. Additional information on the disaster is available in [28, 29].

5.7 Mine Accidents: Poland

Mining is an important industry in Poland. Some of the items mined in the country are copper, zinc, coal, silver, sulfur, lead, lime, nitrogen, and salt. In 2000, the mining and quarrying industrial sector (including mineral fuels and processing) accounted for about 6% of Poland's Gross Domestic Product (GDP).

Over the years, a large number of mining accidents have resulted in fatalities in Poland. In fact, during the period 1880–2006, there were over 25 such accidents [30]. Some of these accidents are as follows [30]:

- **Renard Coal Mine Disaster.** This mine is located at Sonsnowiec, Poland, where an accident in 1880 caused 200 fatalities.
- **Kleofas Coal Mine Disaster.** This mine is located at Katowice, Poland, where an accident in 1896 caused 104 fatalities.
- **Rozbark Mine Disaster.** This mine is located in Bytom, Poland, where an accident in 1923 caused 145 fatalities.
- **Waclaw Coal Mine Disaster.** This mine is located at Nowa Ruda, Poland, where an accident in 1931 caused 151 fatalities.
- **Makoszowy Coal Mine Disaster.** This mine is located at Zabrze, Poland, where an accident in 1958 caused 72 fatalities.
- **Silesia Coal Mine Disaster.** This mine is located at Czechowice-Dziedice, Poland, where an accident in 1974 killed 34 persons.
- **Dymitrow Coal Mine Disaster.** This mine is located at Bytom, Poland, where an accident in 1979 caused 34 deaths.
- **Walbrzych Coal Mine Disaster.** This mine is located at Walbrzych, Poland, where an accident in 1985 caused 18 fatalities.
- **Myslowice Coal Mine Disaster.** This mine is located at Myslowice, Poland, where an accident in 1987 killed 18 persons.
- **Miechowice Coal Mine Disaster.** This mine is located at Bytom, Poland, where an accident in 1993 caused 6 fatalities.
- **Jas-Mos Coal Mine Disaster.** This mine is located at Jastrzehie Zdraj, Poland, where an accident in 2002 killed 10 persons.
- **Halemba Colliery Disaster.** This colliery is located at Ruda Slaska, Poland, where an accident in 2006 caused 23 fatalities.

Additional information on mining disasters in Poland is available in [30].

5.8 Mine Accidents: Russia and the Ukraine

Russia and the Ukraine are two of the leading mineral producing countries in the world. Russia produces minerals such as palladium, nickel, aluminum, potash, copper, gold, coal, and lead. Similarly, the Ukraine produces minerals such as coal, iron ore, manganese, and natural gas. A large number of people are employed in the mining industrial sector in both countries. For example, in 2002, over one million people were employed in the mineral extraction sector in Russia, and the Ukraine employs around half million people in its coal industry alone [31].

Over the years, many mining accidents have resulted in fatalities in both Russia and the Ukraine. Some of these were the Ulyanovskaya coal mine disaster, the Zasyadko coal mine disaster, the Barakova Coal mine disaster, the Zyryanouskaya coal mine disaster, the Skochynksy coal mine disaster, and the Donbass coal mine disaster. These disasters occurred in 2007, 2007, 2000, 1997, 1998, and 2004, respectively. Two of these disasters are described below, separately.

5.8.1 Ulyanovskya Coal Mine Disaster

This mine is located at Novokuznetsk, Kuzbass Siberia, Russia. The disaster occurred on March 19, 2007 and it caused 108 fatalities. A subsequent investigation into the disaster revealed that it was caused by a methane explosion [32]. Furthermore, the explosion was caused by sparks from an exposed cable igniting methane gas, which in turn ignited coal dust in the surrounding area. Additional information on the disaster is available in [32].

5.8.2 Zasyadko Coal Mine Disaster

This coal mine is located at Donetsk, Ukraine. On November 18, 2007, an accident in this mine caused 101 fatalities. The accident was the worst mining disaster in the history of the Ukraine. A subsequent investigation into the disaster revealed that the cause of the accident was a methane explosion 1000 m below ground level. Additional information on the disaster is available in [33].

5.9 Mine Accidents: South Africa

South Africa is one of the leading mining nations in the world. It's mining industry in 2007 employed over 490,000 workers and contributed around 18% to nation's Gross Domestic Product (GDP). South Africa is the world's largest producer of platinum, manganese, vanadium, chrome, and vermiculite, and second largest producer of palladium, rutile, zirconium, and ilmenite [34]. Furthermore, the country is still second in gold production in the world, and is also the world's third largest coal exporter.

Over the years, many mining accidents have resulted in fatalities in South Africa. In fact, according to [35], since the turn of the 20th century over 60,000 miners have died and more than a million have been seriously injured in mining acci-

Table 5.3 Decade breakdowns of disasters and fatalities in South African mines from 1900–1989

No.	Time Period: decade	No. of mine disasters	Fatalities in mine disasters
1	1900–1909	34	718
2	1910–1919	51	464
3	1920–1929	31	461
4	1930–1939	48	615
5	1940–1949	31	425
6	1950–1959	51	465
7	1960–1969	48	1041
8	1970–1979	46	539
9	1980–1989	54	834

dents. Furthermore, a South African who spends twenty years working in underground mines faces approximately a one in thirty chance of dying in an occupational accident [35].

Some of the major disasters that have occurred in South African mines are the Coalbrook coal mine disaster in 1960 [35], the Kinross gold mine disaster in 1986, the Hlobane mine disaster in 1983, the St. Helena mine disaster in 1987, and the Secunda coal mine disaster in 1993 [36]. In these five disasters, there were 437, 177, 68, 63, and 53 fatalities, respectively.

The decade breakdowns of number of disasters and fatalities that have occurred in South African mines from 1900–1989 are presented in Table 5.3 [36].

It should be noted from Table 5.3 that the highest number of fatalities (1041) in South African mines occurred in the 1960s and the second highest fatalities (834) in the 1980s.

5.10 Problems

1. Write an essay on global mine accidents.
2. What were the factors responsible for the safety-related gains in the United States mining industry?
3. Discuss the following mining disasters:
   - Monongah mining disaster
   - Cherry mine disaster
4. Compare United Kingdom and United States mining accidents.
5. What were Wales' four worst mining accidents?
6. Compare Chinese and United States mining disasters.
7. Discuss mining accidents in Australia.

8. Discuss the following mining disasters:
   - Hillcrest mining disaster
   - Springhill mining disaster
   - Westray mining disaster
9. Write an essay on Polish mining disasters.
10. Compare Russian and Ukranian mining disasters to South African mining disasters.

References

1. Terazawa K, Takatori T, Tomii S, Nakano K. Methane Asphyxia: Coal Mine Accident Investigation of Distribution of Gas. Am J Forensic Med Pathol 1985; 6(3):211–214.
2. Kucuker H. Occupational Fatalities Among Coal Mine Workers in Zonguldak, Turkey, 1994–2003. Occup Med 2006;56(2): 144–146.
3. Historical Data on Mine Disasters in the United States. United States Department of Labor. Washington, D.C.; 2000.
4. Honkeiko Colliery Mining Disaster. London: Encyclopedia Britannica; 2009. Available from: URL: www.britannica.com/EBchecked/topic/1503377/Honkeiko-colliery-mining-disaster.
5. Konljenovic D, Groves WA, Kecojevic VJ. Injuries in U.S. Mining Operations-A Preliminary Risk Analysis. Safety Science 2008; 46;792–801.
6. Coal Mines: Bureau of Mines Bulletin No. 686, Historical Summary of Coal Mine Explosions in the United States, 1810–1958. Washington, D.C.: Bureau of Mines; 1960.
7. Injury Trends in Mining. Mine Safety and Health Administration (MSHA). 1100 Wilson Boulevard, 21st Floor, Arlington, Virginia; 2009. Available from: URL: www.msha.gov/MSHAINFO/FactSheets/MSHAFCT2.HTM.
8. All Mining Fatalities by State. Mine Safety and Health Administration (MSHA). 1100 Wilson Boulevard, 21st Floor, Arlington, Virginia; 2009. Available from: URL: www.msha.gov/stats/charts/allstatesnew.asp.
9. Monongah Mining Disaster of 1907. London: Encyclopaedia Britannica; 2009. Available from: URL: www.britannica.com/EBchecked/topic/389922/Monongah-mining-disaster-of-1907.
10. Monongah Mining Disaster. Wikipedia 2009. Available from: URL: http://en.wikipedia.org/wiki/Monongah-Mining-Disaster.
11. Briggs EF. A Fairer Estimate of the Victims of the Monongah Disaster Would be Upward of 500. Science 1964;146(2):1–2.
12. Monongah Mine Disaster. West Virginia Division of Culture and History. West Virginia, USA: West Virginia Archives and History; 2009.
13. Stout S. Story of the Great Cherry Coal Mine Disaster. Journal of the Illinois State Historical Society 1979; LXII(1):3–5.
14. Tintory K. Trapped: The 1909 Cherry Mine Disaster (Illinois). Memphis, Tennessee: Atria Publishing; 2002.
15. Mining Accident. Wikipedia 2009. Available from: URL: http://en.wikipedia.org/wiki/Mining-accident.
16. Brown JH. The Valley of the Shadow: An Account of Britain's Worst Mining Disaster, the Senghenydd Explosion. Port Talbot, U.K.: Alun Books; 1981.

17. Phillips JB. Senghenydd: A Brave Community. Abertillery, U.K.: The Old Bakehouse Publications; 2002.
18. Duckham H, Duckham BF, Great Pit Disasters: Great Britain 1700 to the Present Day. London: David and Charles; 1973.
19. Honkeiko Colliery Mining Disaster. London: Encyclopaedia Britannica; 2009. Available from: URL: www.britannica.com/EBchecked/topic/1503377/Honkeiko-colliery-mining-disaster.
20. Sunjiawan Mine Disaster. Wikipedia 2009. Available from: URL: http://en.wikipedia.org/wiki/2005-Sunjiawan-mine-disaster.
21. Mining in Australia. Wikipedia 2009. Available from: URL: http://en.wikipedia.org/wiki/Mining-in-Australia.
22. Piggin S, Lee H. The Mount Kembla Disaster. Melbourne, Australia,: Oxford University Press; 1992.
23. Barwick J. Australia's Worst Disasters: Mining Disasters. Port Melbourne, Victoria, Australia: Heinemann; 1999.
24. Anderson FW. Canada's Worst Mine Disaster. Calgary, Canada: Frontier Books; 1969.
25. Hillcrest Mine Disaster. Wikipedia 2009. Available from: URL: http://en.wikipedia.org/wiki/Hillcrest-mine-disaster.
26. Springhill Mining Disaster. Wikipedia 2009. Available from: URL: http://en.wikipedia.org/wiki/Springhill_Mining_Disaster.
27. 1887 Nanaimo Mine Explosion. Wikipedia 2009. Available from: URL: http://en.wikipedia.Nanaimo-mine-explosion.
28. Westray Mine. Wikipedia 2009. Available from: URL: http://en.wikipedia.org/wiki/Westray-Mine.
29. Amyotte PR, Oehmen AM. Application of a Loss Causation Model to the Westray Mine Explosion. Trans Of the Institution of Chemical Engineers 2002; 80(Part B, January):55–59.
30. List of Mining Disasters in Poland. Wikipedia 2009. Available from: URL: http://en.wikipedia.org/wiki/List_of_mining_disasters_in_Poland.
31. Mining Industry of Russia. Wikipedia 2009. Available from: URL: http://en.wikipedia.org/wiki/Mining-in-Russia.
32. Ulyanovskaya Mine Disaster. Wikipedia 2009. Available from: URL: http://en.wikipedia.org/wiki/Ulyanovskaya-Mine-disaster.
33. Zasyadko Mine Disaster. Wikipedia 2009. Available from: URL: http://en.wikipedia.org/wiki/w2007-Zasyadko-mine-disaster.
34. Mining Industry of South Africa. Wikipedia 2009. Available from: URL: http://en.wikipedia.org/wiki/Mining-in-South _Africa.
35. Van Der Merwe JN. Beyond Coalbrook: What Did We Really Learn?. The Journal of the South African Institute of Mining and Metallurgy 2006;106:857–868.
36. Leger JP. Trends and Causes of Fatalities in South African Mines. Safety Science 1991; 14:169–185.

Chapter 6
Human Factors and Error in Mine Safety

6.1 Introduction

Human factors is an important field whose reason for existence is the various types of errors that people make in performing their tasks. The history of the human factors field may be traced back to 1898, when Frederick W. Taylor conducted various studies to find the most effective designs for shovels [1]. By 1945, human factors engineering was recognized as a specialized subdiscipline. In 1958, H.L. Williams pointed out that human reliability must be factored into overall system reliability predictions or else such predictions would be unrealistic [2].

Human factors and error in the area of safety have become an important issue because each day approximately 9,000 workers sustain injuries on the job and about 140 workers die from work-related illnesses in the United States alone [3, 4].

The first formal human factors-related study in the area of mining was performed in 1971 [5, 6]. This study was specifically concerned with identifying human factors problems in underground coal mines. In 1982, two studies concerning the identification of human factors problems in surface mining were conducted: one dealt with the mining process itself and the other with the processing plants [7, 8]. These three studies directly or indirectly considered human error and safety.

Over the years many publications on various aspects of human factors and error in mine safety have appeared in published literature [6, 9, 10]. This chapter presents various important aspects of human factors and error that are either directly or indirectly concerned with mine safety.

6.2 The Need for the Application of Human Factors to Mining Industries and Common Roadblocks in Introducing Human Factors to Organizations

In today's competitive global environments, technology is playing an important role in increasing productivity and safety in the area of mining. This, in turn, has created a definite need for paying close attention to determining how humans and technology can work together in an effective manner. More specifically, how should mining tasks and equipment be designed to match the limitations and capabilities of the mine workers who will be performing such tasks and operating and maintaining such equipment in the field environment [6, 10]?

The application of human factors principles to mining industries can be very useful to address questions such as this. Past experience indicates, however, that there are many roadblocks to the application of human factors in many organizations, including the area of mining, that arise from misunderstandings. Some of the common ones are as follows [6, 10]:

- **All humans are created equal.** To this notion it can be added that no two humans are absolutely identical. Thus, systems must be designed in such a way that they are able to effectively accommodate the diversity of the human population.
- **Humans can be trained to overcome design-related shortcomings.** Although this is true to a certain degree, training can also be somewhat unreliable. More specifically, experience indicates that in stressful conditions humans generally respond the way they think systems should function, which may not be how the systems actually function or operate.
- **Engineers and designers know best how humans think and act.** This may be true to a certain degree, but engineers and designers do not necessarily act or think like other members of the society-at-large.
- **Minor human factors shortcomings are not important.** Experience indicates that minor shortcomings often compound into major ones and normally have a way of insidiously eating into productivity and efficiency.
- **No serious incidents indicate no human factors-associated problems.** It is to be noted that although the Three Mile Island nuclear power generating station had various human factors-associated shortcomings in its control room, it reported no serious incidents whatsoever until the day it came within an hour of meltdown [6, 10].

6.3 Occupational Stressors and Human Factors-related Formulas

There are many occupational stressors and they may be classified under the following four categories [4, 11]:

- **Workload-related stressors.** These stressors are concerned with the problems of work overload or underload. In the former case, the task or job requirements exceed the capability of a person. In the latter case, the work performed by a person fails to provide them with sufficient stimulation. Two examples of work underload are repetitive performance of the same task and lack of opportunity to use an individual's acquired skills and knowledge.

- **Occupational change-related stressors**. These are concerned with an individual's cognitive, physiological, and behavioral patterns of functioning, and they usually exist in organizations concerned with productivity and growth. Three examples of occupational change are organizational restructuring, relocation, and promotion.
- **Occupational frustration-related stressors.** These stressors result in situations where the assigned task/job inhibits the fulfillment of stated goals or objectives. Some of the factors that form occupational frustration elements are lack of communication, role ambiguity, and bureaucracy-related difficulties.
- **Miscellaneous stressors.** These stressors include poor interpersonal relationships, noise, and too little or too much lighting.

There are many mathematical formulas in the published literature to estimate human factors-related information. Some of these formulas considered useful for application in the area of mine safety are presented below [10, 12, 13].

6.3.1 Maximum Lifting Load Estimation Formula

This formula is concerned with estimating the maximum lifting load for a person. The maximum lifting load is expressed by [14]

$$MLL_p = \theta(MS_{ib}) \tag{6.1}$$

where

MLL_p	is the maximum lifting load for a person.
MS_{ib}	is the isometric back muscle strength.
θ	is a constant whose values for females and males are 0.95 and 1.1, respectively.

6.3.2 Rest Period Length Estimation Formula

This formula is concerned with estimating the length of required rest periods for humans carrying out various types of tasks. The length of a rest period is expressed by [10, 15]

$$RRPL = WT(AEE - SEE)/(AEE - RL) \tag{6.2}$$

where

$RRPL$	is the required period length, expressed in minutes.
WT	is the working time, expressed in minutes.
RL	is the resting level. Usually, its value is taken as 1.5 kcal/min.

SEE is the standard energy expenditure expressed in kilocalories per minute (kcal/min). Its value is taken to be 5 kcal/min, when there are no data.

AEE is the average energy expenditure expressed in kcal/min of work.

Example 6.1

Assume that a person is performing a mining task for 70 min and his/her average energy expenditure is 5 kilocalories per minute. Calculate the length of the required rest period if the standard energy expenditure is 4 kilocalories per minute.

By substituting the specified data values into Equation 6.2, we get

$$RRPL = 70(5-4)/(5-1.5) = 20\,\text{min}$$

Thus, the length of the required rest period is 20 min.

6.3.3 Character Height Estimation Formula

The ability of humans to visually discriminate objects depends specifically on factors such as illumination, size, and exposure time. This formula is concerned with determining character height by taking into consideration factors such as viewing distance and conditions, the importance of reading accuracy, and illumination. Thus, character height is defined by [10, 16]

$$CH = \alpha VD + CF_c + CF_{iv} \tag{6.3}$$

where

CH is the character height expressed in inches.

α is a constant whose value is taken as 0.0022.

CF_c is the correction factor for criticality or importance. Its recommended values for critical and non-critical markings are 0.075 and 0, respectively.

VD is the viewing distance expressed in inches.

CF_{iv} is the correction factor for illumination and viewing conditions. Its recommended values are available in [17].

6.4 Human Factors-related Considerations in Mining Equipment Design

In order to produce human-compatible and safe mining equipment, it is important to consider relevant human factors during the mining equipment design stage. The main objective of this exercise is to insure that the equipment does not subject humans to extreme physical or mental stress or hazards, and allows them to carry out their tasks in the most effective manner.

Human factors-related considerations during the following four stages of the mining equipment design are as follows [13, 18]:

- **Pre-conceptual stage.** At this stage, design professionals should systematically define items such as the operational and mission requirements, the performance-related requirements for each and every mission, and the functions required for performing each mission event.
- **Conceptual stage.** At this stage, the design professionals should include items such as a preliminary definition of manning and training requirements, analysis to define the most appropriate design approach for accomplishing each hardware functional assignment, and preliminary task descriptions of all potential users, maintainers, and operators.
- **Pre-design stage.** At this stage, the design professionals should review the conceptual stage analysis and perform time line and link analyses, man-machine mock-up studies, etc.
- **Detailed-design stage.** At this stage, the design professionals should develop an equipment/system statement, perform link analyses for all important man-machine interfaces, identify critical skill requirement specifications, create and evaluate critical man-machine mock-ups, etc.

6.5 Human Factors Safety Issues

There are many human factors safety issues that must be considered early in a safety program because they may directly or indirectly affect safety. Some of these issues concern safety and health, displays and controls, training, visual/oral alerts, anthropometrics, communications, environment, special skills and tools, information requirements, and work space [4, 19].

The safety and health issues are concerned with preventing exposure to safety and health hazards for operators, maintainers, and others. The displays and controls issues are concerned with designing and arranging displays and controls with respect to a natural sequence of operational actions for operators and maintainers. The training issues are concerned with reducing or minimizing the requirements for responses of operators, maintainers, and others.

The visual/oral alerts issues are concerned with effective design of auditory and visual alerts for invoking appropriate responses of operators, maintainers, and others. The anthropometrics issues are concerned with accommodating humans represented in the user population (i.e., from the 5th through 95th percentile levels of the physical characteristics) in the system design. The communications issues are concerned with system design considerations that enhance required user teamwork and communications.

The environment issues are concerned with accommodating environment-related factors to which humans or systems will be subjected. The special skills and tools issues are concerned with reducing the requirement for special operator/maintainer abilities, tools, skills, or characteristics. The information requirements issues are concerned with insuring the availability of information needed by individuals such as maintainers and operators for performing certain tasks or jobs. In particular, the information should be available both when it is needed and in the proper sequence. Finally, the work space issues are concerned with the existence of sufficient work space for humans and for their equipment and tools.

6.6 Classifications and Causes of Human Errors Resulting in Fatal Mine Accidents

Experience indicates that there are various types of human errors that result in fatal mine accidents. They occur due to various causes and may be grouped under various classifications. For example, a study of 794 human errors that resulted in fatal accidents in mines grouped these errors under six classifications [20]. These classifications and their causes are presented below [10, 20]:

- **Failure to perceive a warning.** Some of the causes for the occurrence of this type of human error were lack of experience, inadequate information, and lack of training.
- **Failure to respond to a recognized warning.** The main cause for the occurrence of this type of human error was underestimation of the hazard.
- **Failure to recognize a perceived warning.** Some of the causes for the occurrence of this type of human error were poor inspection methods, inattention or distraction, neglect of proper inspections, masking noise, and obstruction of the line of sight.
- **Ineffective response to a warning.** Three main causes for the occurrence of this type of human error were carelessness or negligence, well-intended but ineffective direct action, and inappropriate standard practice.
- **Underestimation of a hazard.** Causes for the occurrence of this type of human error were unidentifiable.
- **Inappropriate secondary warning.** Causes for the occurrence of this type of human error were also unidentifiable.

6.7 Common Mining Equipment Maintenance Errors and Maintenance Error Contributory Factors

There are many mining equipment maintenance errors that may compromise safety. Some of the common ones are presented in Table 6.1 [10, 21].

Table 6.1 Common mining equipment maintenance errors

No.	Common Maintenance Error
1	Failure to detect while inspecting
2	Use of incorrect fluids, lubricants, or greases
3	Failure to follow prescribed procedures and instructions
4	Omitting a part or component
5	Failure to close or seal properly
6	Failure to lubricate
7	Installation of incorrect part
8	Reassemble error
9	Failure to act on indicators of problems due to factors such as workload, time constraints, or priorities
10	Failure to check, align, or calibrate
11	Error resulting from failure to complete task due to shift change
12	Parts or components installed backwards

There are many factors that contribute to mining equipment maintenance errors. Some of the important ones are shown in Figure 6.1 [10, 21].

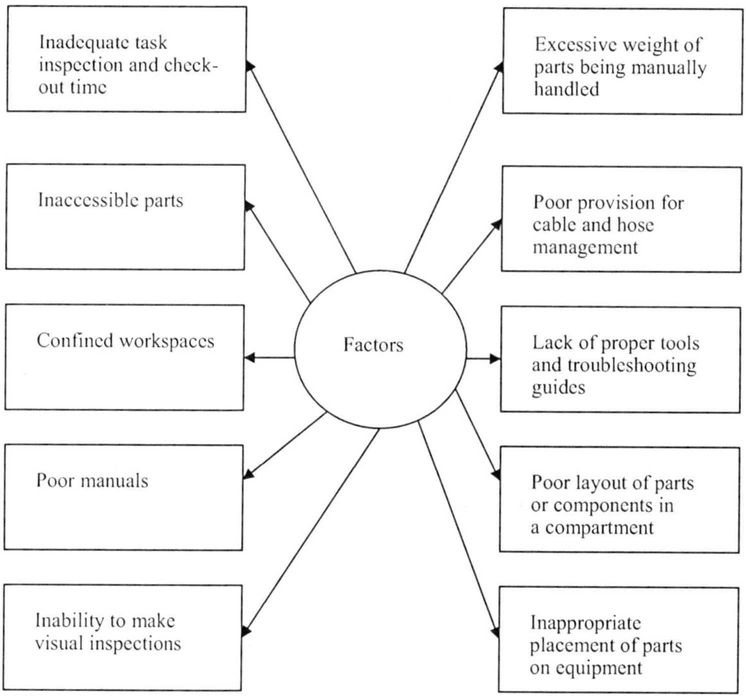

Figure 6.1 Important mining equipment maintenance error contributory factors

6.8 Types of Chemicals Released in Human Error-associated Events in the Mining and Manufacturing Industrial Sectors

Experience indicates that human error has been an important factor in the occurrence of large scale hazardous material-related events. For example, a study of over three thousand human error-associated events in the mining and manufacturing industrial sectors during the period 1996–2003 reported that various types of chemicals were released [22]. Most of these chemicals are shown in Figure 6.2 [22].

It should be noted that although the number of chemicals released per human error-associated event ranged from 1 to 14, in most of these events only one chemical was released.

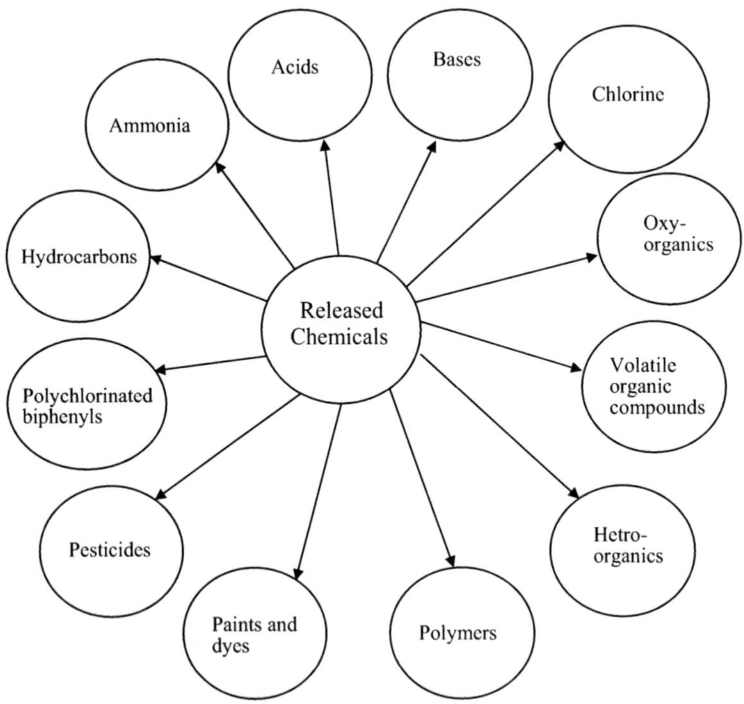

Figure 6.2 Types of chemicals released during human error-associated events in mining and manufacturing industries

6.9 Methods for Performing Human Error Analysis in the Area of Mine Safety

There are many methods and techniques that can be used to perform human error analysis of engineering systems [23]. Most of these methods and techniques can also be used to perform human error analysis in the area of mine safety. Three of these methods and techniques considered most useful for application in the area of mine safety are described below.

6.9.1 Fault Tree Analysis

This method is described in Chapter 4 and its application to human error analysis in the area of mine safety is demonstrated through the following example.

Example 6.2

Assume that a mine maintenance worker can commit an error due to five factors and the error can cause a serious accident. The five factors are poor environment, inadequate training, inadequate tools, carelessness, or poor instructions. Two principal reasons for the poor instructions are poorly written maintenance procedures or poor verbal instructions.

Develop a fault tree for the top event "Mine maintenance worker committed an error" by using the fault tree symbols given in Chapter 4, and calculate the probability of the occurrence of the top event if the probabilities of occurrence of independent events poor environment, inadequate training, inadequate tools, carelessness, poorly written maintenance procedures, and poor verbal instructions are 0.07, 0.06, 0.05, 0.04, 0.03, and 0.02, respectively.

A fault tree for the example is shown in Figure 6.3. The single capital letters in the fault tree demote corresponding events (e.g., X for inadequate tools and I for poor instructions).

Using an equation given in Chapter 4 for the probability of occurrence of the OR gate output fault event and the given data values, the probability of occurrence of event I in Figure 6.3 is

$$\begin{aligned} P(I) &= 1-(1-P(Z))\,(1-P(L)) \\ &= 1-(1-0.02)\,(1-0.03) \\ &= 0.0494 \end{aligned}$$

where

$P(Z)$ is the probability of occurrence of event Z.
$P(L)$ is the probability of occurrence of event L.

Using the calculated value above, the given data values, and an equation given in Chapter 4 for the probability of occurrence of the OR gate output fault event, the probability of occurrence of top event T in Figure 6.3 is

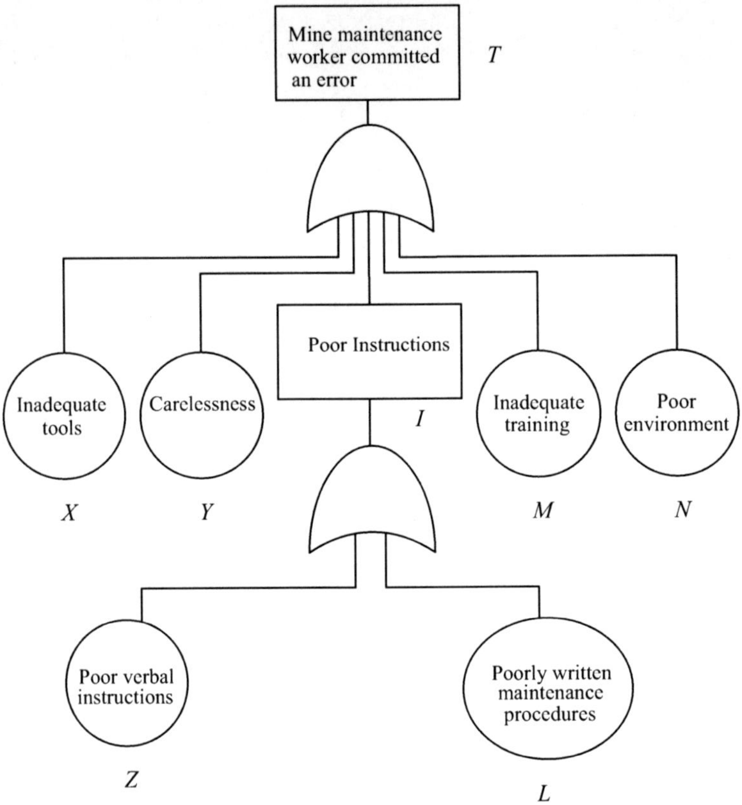

Figure 6.3 A fault tree for Example 6.2

$$
\begin{aligned}
P(T) &= 1-(1-P(X))\,(1-P(Y))\,(1-P(I))\,(1-P(M))\,(1-P(N)) \\
&= 1-(1-0.05)\,(1-0.04)\,(1-0.0494)\,(1-0.06)\,(1-0.07) \\
&= 0.2421
\end{aligned}
$$

where

$P(X)$ is the probability of occurrence of event X.
$P(Y)$ is the probability of occurrence of event Y.
$P(M)$ is the probability of occurrence of event M.
$P(N)$ is the probability of occurrence of event N.

Thus, the probability of the occurrence of the top event T (i.e., mine maintenance worker committed an error) is 0.2421.

6.9.2 Throughput Ratio Method

This is a useful method developed by the United States Navy Electronic Laboratory Center [24]. The ratio generated by this approach determines the man-machine interface operability. The term "throughput" implies transmission because the ratio is expressed in terms of responses per unit time emitted by the system or equipment operator. The throughput ratio in percentage is defined by [23, 24]

$$MMO = \left[\frac{\mu}{\alpha} - f_c \right](100) \tag{6.4}$$

where

- MMO is the man-machine operability expressed as a percentage.
- A is the total number of throughput items to be generated per unit time to satisfy design-related expectations.
- μ is the total number of throughput items generated per unit time.
- f_c is the correction factor (i.e., out-of-tolerance output or correction for error).

The correction factor, f_c, is defined by

$$f_c = \left[\left(\frac{\mu}{\alpha}\right)\left(\frac{m_l}{m_2}\right) \right] \left[\left(\frac{\mu}{\alpha}\right)\left(\frac{m_l}{m_2}\right) P_{fh} P_{nde}^2 \right] \tag{6.5}$$

where

- P_{fh} is the probability of function failure due to human error.
- P_{nde} is the probability that the equipment operator will not be able to detect an error.
- m_1 is the total number of trials in which the control-display operation is carried out incorrectly.
- m_2 is the total number of trials in which the control-display operation is carried out.

This approach may be used in the area of mine safety for purposes such as establishing system feasibility, making comparisons of alternative design operabilities, and demonstrating system acceptability.

Example 6.3

For the following specified values of P_{fh}, P_{nde}, m_1, m_2, α, and μ, calculate the value of the throughput ratio by using Equations 6.4 and 6.5:

$$P_{fh} = 0.7$$
$$P_{ndc} = 0.2$$
$$m_1 = 5$$
$$m_2 = 20$$
$$a = 12$$
$$\mu = 6$$

By substituting the values specified above into Equation 6.5, we get

$$f_c = \left[\left(\frac{6}{12}\right)\left(\frac{5}{20}\right)\right]\left[\left(\frac{6}{12}\right)\left(\frac{5}{20}\right)(0.7)(0.2^2)\right]$$
$$= 0.0004375$$

By substituting the value calculated above and the specified data values into Equation 6.4, we obtain

$$MMO = \left[\frac{6}{12} - 0.0004375\right](100)$$
$$= 49.96\%$$

Thus, the calculated value of the throughput ratio (i.e., man-machine operability) is 49.96%.

6.9.3 Probability Tree Method

This is a useful method that is concerned with conducting task analysis by diagrammatically representing all critical human actions and other events associated with the system under consideration. The probability tree branches denote diagrammatic task analysis. More specifically, the branching limbs of the tree denote outcomes (i.e., success or failure) of each event, and each branching limb is assigned an occurrence probability.

Some of the principal advantages of the probability tree method are as follows [23]:

- A powerful visibility tool
- An useful tool to incorporate, with some modifications, factors such as interaction stress, interaction effects, and emotional stress
- An useful tool in applying predictions of individual error rates and prediction of the quantitative effects of errors
- An useful tool for reducing the probability of errors due to computation caused by computational simplification

The following example demonstrates the application of this method to a mine safety-related problem.

Example 6.4

Assume that a mine worker performs three consecutive tasks i, j, and k and each can either be performed correctly or incorrectly. The incorrect performance of any of these tasks can result in an accident. Task i is performed before task j and in turn task j is performed before task k, and all these tasks are independent of each other.

Develop a probability tree and obtain a probability expression that the mine worker will not successfully accomplish the overall mission.

A probability tree for this example is shown in Figure 6.4. The probability tree in the figure shows that the mine worker first performs task *i* correctly or incorrectly, then proceeds to task *j*, which can also be performed correctly or incorrectly, and then finally proceeds to task *k* which can also be performed correctly or incorrectly.

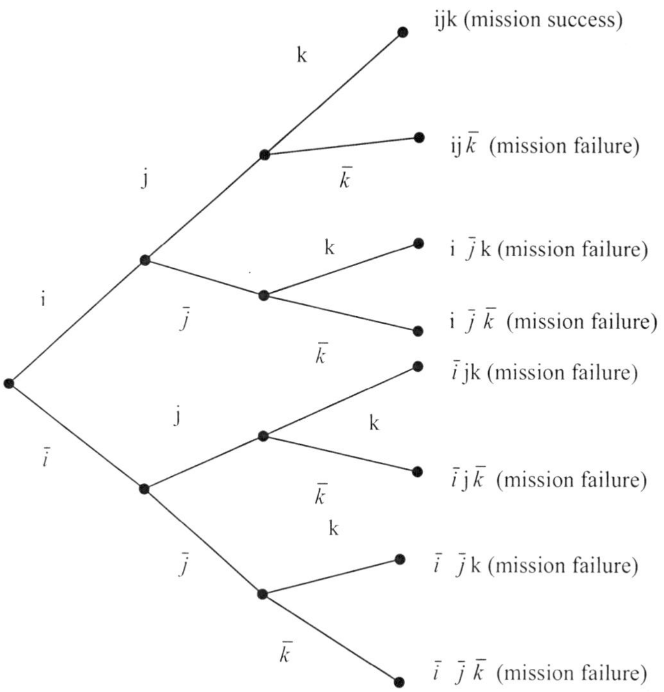

Figure 6.4 Probability tree diagram for Example 6.4

In Figure 6.4, single letters *i*, *j*, and *k* with bars denote unsuccessful events (i.e., tasks *i*, *j*, and *k* performed incorrectly) and without bars denote successful events (i.e., tasks *i*, *j*, and *k* performed correctly). Other symbols utilized for obtaining the probability expression for the example are defined below:

P_i is the probability of performing task *i* correctly.
P_j is the probability of performing task *j* correctly.
P_k is the probability of performing task *k* correctly.
$P_{\bar{i}}$ is the probability of performing task *i* incorrectly.
$P_{\bar{j}}$ is the probability of performing task *j* incorrectly.
$P_{\bar{k}}$ is the probability of performing task *k* incorrectly.
P_{ns} is the probability of not successfully accomplishing the overall mission by the mine worker.

Using Figure 6.4 and the above symbols, we write down the following probability expression for the mine worker not completing the overall mission successfully:

$$P_{ns} = P_i P_j P_{\bar{k}} + P_i P_{\bar{j}} P_k + P_i P_{\bar{j}} P_{\bar{k}} + P_{\bar{i}} P_j P_k + P_{\bar{i}} P_j P_{\bar{k}} + P_{\bar{i}} P_{\bar{j}} P_k + P_{\bar{i}} P_{\bar{j}} P_{\bar{k}} \quad (6.6)$$

Thus, the expression for the probability that the mine worker will not accomplish the overall mission successfully is given by Equation 6.6.

6.10 Design Improvement Guidelines for Reducing Mining Equipment Maintenance Errors and Factors Responsible for Failing to Reduce the Occurrence of Human Error in the Mining Sector

Over the years engineering professionals working in the area of mining have developed various engineering design improvement guidelines for reducing mining equipment maintenance errors. Some of these guidelines are as follows [21]:

- Aim to design to facilitate detection of all types of errors
- Use effective operational interlocks so that subsystems cannot be turned on when they are incorrectly installed or assembled
- Improve warning readouts, indicators, and devices to reduce or minimize human decision making
- Improve fault isolation design by indicating the fault direction, designating appropriate test procedures and points, and providing built-in test capability
- Use decision guides for reducing or minimizing human guesswork by providing appropriate arrows to indicate correct hydraulic pressures, correct type of fluids or lubricants, and flow direction
- Improve equipment-part interface by designing interfaces so that the part can only be installed correctly and provide appropriate mounting pins and other devices for supporting a part while it is being unbolted or bolted

There are many factors that are responsible for failing to reduce the occurrence of human error in the mining sector. Four of these factors are as follows [25]:

- The greater degree of automation and mechanization in today's modern mining sector requires greater efficiency, capability, and understanding from all types of mining workers.
- There is greater mental tension and worry among today's mining workers because of a greater desire to have more than others.
- Stress in the home is on the rise throughout the world.
- Today's mining workers are carrying out their assigned tasks under more difficult physical, environmental, and geo-mining conditions than ever before.

6.11 Problems

1. Discuss at least five common roadblocks for introducing human factors to organizations.
2. Discuss the four classifications of occupational stressors.
3. Assume that a mining worker is performing a task for 50 min and his/her average energy expenditure is 4 kilocalories per minute. Calculate the length of the required rest period if the standard energy expenditure is 3 kilocalories per minute.
4. Discuss human factor safety issues.
5. Discuss classifications and causes of human errors that have resulted in fatal mine accidents.
6. List at least ten common mining equipment maintenance errors.
7. What are the main maintenance error contributory factors?
8. What were the types of chemicals released during human error-associated events in the mining and manufacturing industries?
9. Discuss at least five useful guidelines for reducing mining equipment maintenance errors.
10. What are the factors responsible for failing to reduce the occurrence of human error in the mining sector?

References

1. Chapanis A. Man-Machine Engineering. Belmont, California: Wadsworth Publishing Company; 1965.
2. Williams HL. Reliability Evaluation of the Human Component in Human-Machine Systems. Electrical Manufacturing 1958; 5:78–82.
3. National Institute for Occupational Safety (NIOSH). Department of Health and Human Services. Washington, D.C.; 2002.
4. Dhillon BS. Engineering Safety: Fundamentals, Techniques, and Applications. River Edge, New Jersey: World Scientific Publishing; 2004.

5. Hitchcock L, Sanders M, editors. Survey of Human Factors in Underground Bituminous Coal Mining Naval Ammunition Depot. Indiana: Crane; 1971.
6. Sanders MS, Pay JM. Human Factors in Mining, Information Circular No. 9182. Bureau of Mines, United States Department of the Interior. Washington, D.C.; 1991.
7. Conway EJ, Sanders MS. Recommendations for Human Factors Research and Development Projects in Surface Mining. Report No. OFR-211-83. Bureau of Mines, United States Department of the Interior. Washington, D.C.; 1982.
8. Gross B, Schurick J. Hazard Analysis and Safety Economics in Mineral Processing Plants. report. Westlake Village, California: Canyon Research Group; 1982.
9. Dhillon BS. Bibliography of Literature on Mining Equipment Reliability. Microelectronics and Reliability 1986; 26:1131–1138.
10. Dhillon BS. Mining Equipment Reliability, Maintainability, and Safety. London: Springer Inc.; 2008.
11. Beech HR, Burns LE, Sheffield BF. A Behavioral Approach to the Management of Stress. New York: John Wiley and Sons; 1982.
12. Dhillon BS. Engineering Design: A Modern Approach. Chicago: Irwin; 1996.
13. Dhillon BS. Advanced Design Concepts for Engineers. Lancaster, Pennsylvania: Technomic Publishing Company; 1998.
14. Poulsen E, Jorgensen K. Back Muscle Strength, Lifting, and Stooped Working Postures. Applied Ergonomics 1971; 2(3):133–137.
15. Altman JW. *et al.* Guidelines to Design of Mechanical Equipment for Maintainability. report No. ADSD-TR-61-381. United States Air Force (USAF), Aeronautical Systems Division. Dayton, Ohio. August 1961.
16. Peters GA, Adams BB. Three Criteria for Readable Panel Markings. Prod Eng 1959; 30(21):55–57.
17. Huchingson RD. New Horizons for Human Factors in Design. New York: McGraw Hill Book Company; 1981.
18. Woodson WE, Tillman P, Tillman B. Human Factors Design Handbook. New York: McGraw Hill Book Company; 1981.
19. Federal Aviation Administration (FAA). System Safety Handbook. Washington, D.C.; 2000.
20. Lawrence AC. Human Error as a Cause of Accidents in Gold Mining. Journal of Safety Research 1974; 6(2):78–88.
21. Under RL, Conway K. Impact of Maintainability Design on Injury Rates and Maintenance Costs for Underground Mining Equipment. in Improving Safety at Small Underground Mines. compiled by RH Peters. Special Publication No. 18–94. Bureau of Mines. US Department of the Interior. Washington, D.C.; 1994.
22. Ruckart PZ, Burgess PA. Human Error and Time of Occurrence in Hazardous Material Events in Mining and Manufacturing. J Hazard Mater 2007; 142:747–753.
23. Dhillon BS. Human Reliability: With Human Factors. New York: Pergamon Press; 1986.
24. Meister D. Comparative Analysis of Human Reliability Models. Report No. AD 734-432, 1971. Available from the National Technical Information Service (NTIS) Springfield, Virginia.
25. Mohan S, Duarte D. Cognitive Modeling of Underground Miners Response to Accidents. Proceedings of the Annual Reliability and Maintainability Symposium 2006; 51–55.

Chapter 7
Mining Equipment Safety

7.1 Introduction

The type of equipment used in mining, has come a long way since man first used hand held tools made of flint and bone for extracting ores from earth. Currently, a vast sum of money is spent each year to produce various types of mining equipment throughout the world. Some examples of such equipment are crushers, haul trucks, hoist controllers, mine carts, and dragline excavators.

Mining equipment has improved quite dramatically over the years. Today, such equipment has become rather complex and powerful, and a high degree of skill is required for its economical and safe operation and maintenance. In the past, many accidents have occurred involving mining equipment. In order to improve safety in United States mines (including equipment safety), the US Congress passed the Mine Safety and Health Act in 1977.

As a result of this congressional act, an agency called Mine Safety and Health Administration (MSHA) was established by the US Department of Labor. The main goal of this agency includes items such as enforcement of compliance with mine-related safety and health standards, promotion of better health and safety conditions in the United States mines, and reduction of health-related hazards [1]. This chapter presents various important aspects of mining equipment safety.

7.2 Mining Equipment Safety-related Facts and Figures and Mining Equipment-related Fatal Accidents

Some of the facts and figures concerning mining equipment safety are as follows:

- In 2004, around 17% of the injuries in the United States underground coal mines were connected to bolting machine equipment [2].

- During the period 1978–1988, maintenance activities in United States mines accounted for about 34% of all lost-time injuries [3].
- During the period 1990–1999, electricity was the 4th leading cause of fatalities in the United States mining sector [4].
- During the period 1990–1999, 76 injuries were caused by 197 equipment fires in coal mining operations in the United States [5].
- During the period 1983–1990, about 20% of the coal mine-related injuries occurred during equipment maintenance or while using various types of hand tools [6].
- A study by the United States Bureau of Mines (now National Institute for Occupational Safety and Health (NIOSH)) reported that equipment was the primary cause of injury in about 11% of all mining accidents and a secondary causal factor in another 10% of the accidents [7–9].

Each year, many fatalities are caused by various types of mining equipment. For example, during the period 1995–2005, 483 fatalities in the United States mining operations were equipment-related [1], or around 44 fatalities per year. Specific numbers of fatalities for different types of mining equipment are presented in Table 7.1 [10].

The percentage distributions of these fatalities with respect to the type of equipment are as follows: haul truck, 22.36%, conveyor, 9.32%, front-end-loader, 8.49%, miner, 6.21%, dozer, 5.8%, drill, 3.31%, shuttle car, 2.69%, roof bolter, 1.45%, LHD, 1.24%, fork-lift, 1.04%, longwall, 1.04%, hoisting, 0.41%, and miscellaneous equipment, 36.65% [10].

It should be noted from these percentage values that for this time period only four subcategories of equipment, i.e., haul truck, conveyor, front-end loader, and miscellaneous equipment, account for over 75% of all equipment-related fatal accidents.

Table 7.1 Equipment-related fatalities for different types of mining equipment for the period 1995–2005

No.	Equipment type	No. of fatalities
1	Haul truck	108
2	Conveyor	45
3	Front-end loader	41
4	Miner	30
5	Dozer	28
6	Drill	16
7	Shuttle car	13
8	Roof bolter	7
9	LHD	6
10	Fork-lift	5
11	Longwall	5
12	Hoisting	2
13	Miscellaneous equipment	177
14	Total	483

7.3 Equipment Fire-related Mining Accidents, Mining Equipment Fire Ignition Sources, and Strategies for Reducing Mining Equipment Fires

Experience indicates that equipment fires can be quite dangerous to the safety of personnel involved in mining operations, particularly when they occur in the confined areas of underground mines. For example, during the 10 year period of 1990–1999, there were 197 equipment fires in the United States coal mines that resulted in 76 injuries [5]. These fires were divided into four categories by NIOSH for further study, as shown in Figure 7.1, [5].

Under the surface coal mine equipment fires category, there were 140 equipment fires and 56 of them resulted in 56 injuries. Similarly, under the underground equipment fires category, there were 26 equipment fires and 10 of them caused 10 injuries. Also, under the prep plant fires category, there were 17 equipment fires and 6 of them resulted in 7 injuries. Finally, under the surface coal mine equipment fires, there were 14 equipment fires and 4 of them caused 4 injuries.

Over the years, it is noted that there could be many ignition sources for mining equipment fires. Four of the important ones are shown in Figure 7.2 [5].

There are various ways to reduce mining equipment fires and injuries, and some of the better methods or strategies are as follows [5]:

- Conduct equipment fuel, hydraulic and electrical system inspections frequently and thoroughly
- Develop new technologies for emergency engine/pump shutoff, emergency hydraulic line drainage/safeguard system, and fire barriers

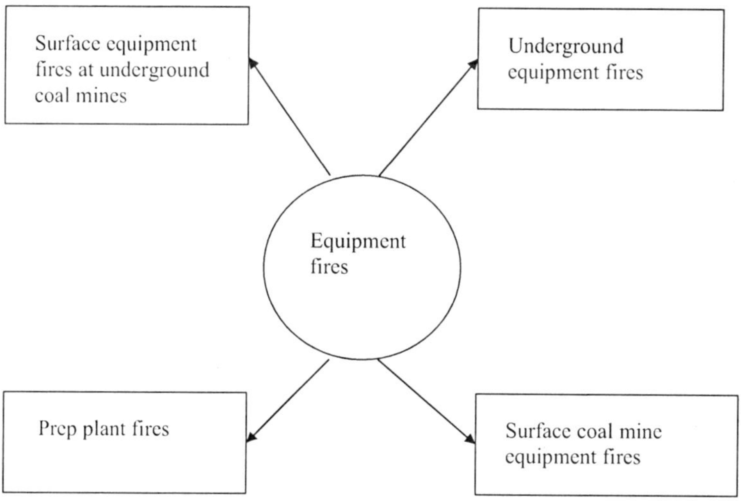

Figure 7.1 Classifications of equipment fires in the United States coal mines during the period 1990–1999

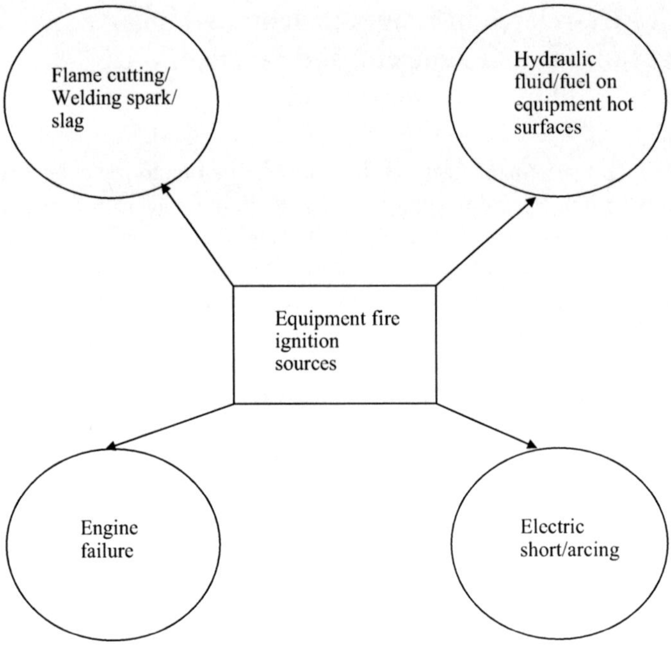

Figure 7.2 Four important mining equipment fire ignition sources

- Improve equipment/cab fire prevention/suppression systems
- Provide frequent and appropriate emergency preparedness training to equipment operators
- Develop effective equipment/cab fire detection systems that have a visible/audible cab alarm

7.4 Fatalities and Injuries Due to Crane, Drill Rig, and Haul Truck Contact with High Tension Power Lines and Guidelines for Improving Electrical Safety in Mines

Overhead or high-tension electric power lines present a very serious electrocution hazard to people working in a variety of industries, because equipment such as cranes, drill rigs, and haul trucks are often exposed to these lines. When contacting the power lines, this equipment becomes elevated to a high voltage, and simultaneous contact to the "hot" frame and ground by humans can result in dangerous electric shocks and burns.

Some of the industries where risk of such accidents is greatest are mining, construction, and agriculture. Each year about 2300 accidental overhead power line contacts occur in the United States [11]. During the period 1980–1997, the United

States mining industry reported at least 94 mobile equipment overhead line contact accidents [11]. These accidents caused 114 injuries and around 33% of them were fatal. Most of these accidents involved cranes (47%), dump bed trucks (24%), and drills (14%) [11].

Over the years various guidelines have been proposed to improve general electrical safety in the mining industry, and these include equipment. Some of these guidelines considered most useful are as follows [4, 12]:

- Use effective power line avoidance devices as much as possible
- Improve system/equipment design in general and with respect to electrical safety in particular
- Provide proper power line awareness training to all concerned individuals
- Make effective improvements in electrical maintenance procedures and schedules
- Target effective training at identified problem areas

7.5 Mining Ascending Elevator Accidents and Safety in Electrical Design of Mine Elevator Control Systems

Many ascending elevator accidents in mining and other areas have occurred over the years. In a number of cases, they have caused serious injuries or fatalities. These accidents have occurred on counter-weighted elevators because of mechanical, structural, and electrical failures. Although the elevator cars contain safe ties that grip the guide rails and stop a falling car, such devices fail to provide proper protection in the upward direction.

In 1987, an ascending elevator car accident at a Western Pennsylvania coal mine caused extensive structural damage and disabled the elevator for a number of months. As a result of this accident, the Pennsylvania Bureau of Deep Mine Safety established an advisory committee for evaluating devices on the market in order to provide effective ascending car overspeed protection for current and new mine elevator installations.

The committee recommended the following four protective methods, in addition to requirements that new elevators have a governor rope monitor device, back out of over travel switch, and a manual reset [13]:

- Counterweight safe ties
- Dynamic braking
- Rope brake (i.e., a pneumatic rope brake that grips the suspension ropes).
- Weight balancing (counterweight equals the empty car weight).

Additional information on the above four methods is available in [13–15].

Safety in the electrical design of mine elevator control systems is very important. For example, an ascending elevator overspeed accident at a Southern Ohio underground coal mine in 1994 caused four miners in the elevator various types of serious injuries, including head lacerations and contusions and fractured vertebrae and ribs [16]. A subsequent investigation into the accident has identified various elevator control system components whose failure could lead to serious elevator accidents.

Some useful design guidelines to reduce the chances of a hoisting accident by improving the critical elevator control circuits and devices for safety integrity and trouble-shooting are as follows [16]:

- Improve safety circuit integrity
- Introduce redundancy when microprocessor or program logic controller (PLC) elevator control is used
- Insure that the electrical brake control design is rugged, reliable, and fault tolerant
- Use an effective governor-rope monitoring system
- Use an effective fault detection mechanism
- Install pit-water alarm to indicate the accumulation of water on the pit floor

7.6 Human Factor Considerations in the Design of Mine Safety and Rescue Devices and Human Factor Tips for Safer Mining Equipment

There are many types of mine safety and rescue devices. They may be divided into two categories: individual safety devices for miners and rescuers, and detecting instruments and rescue devices for rescue squads. Some examples of individual safety devices are miner caps, cap lamps, oxygen breathing apparatus for rescuers, and portable self-rescuing boxes for miners. Similarly, some examples of detecting instruments and rescue devices for rescue squads are gas detectors for detecting dangerous gases, portable rescuing tools, drillers, and pumps.

Some important human factors-related design considerations for the effective performance of mine safety and rescue devices are as follows [17]:

- **Miners' physiological factors.** These include items such as height, sight, weight, the force that limbs and trunk can provide, and color identification capabilities.
- **Miners' psychological factors.** These include items such as emotional behavior; psychological reaction in emergency situations, etc.
- Miners' behavior under different production environments.
- Miners' degree of professional training.

Some of the human factors-related tips for safer mining equipment are as follows [12, 18]:

- Insure that all operators can identify all of the necessary controls accurately and quickly
- Insure that the workstation effectively fits potential operators from the 5th to 95th percentile range

- Anticipate potential safety-related hazards and necessary emergency actions before starting the design process
- Design the seat in such a way that miners can replace or maintain it with ease
- Insure that there is sufficient contrast between object luminance or location of interest and the surrounding background, so that a given task can be performed safely
- Insure that all controls have adequate resistance for reducing the chance of inadvertent activation by the weight of a foot or hand
- Aim to distribute workloads as evenly as possible between feet and hands
- Insure that each and every design control effectively complies with anthropometric data concerning human operators
- Insure that seats do not hinder the ability of an operator to control the machine or equipment
- Insure that each and every design control can withstand/guard against abuse (e.g., from falling roofs or from the forces imposed during a panic response in emergency circumstances)
- Insure that seats adjust and effectively fit to body dimensions, distribute weight for relieving pressure points, and support posture
- Insure that the work station provides an unobstructed line of sight to objects/locations that should be clearly visible to perform a given task safely
- Insure that seats do not hinder the operator's ability when entering or exiting the workstation
- Insure that the relative placement of displays and controls for similar machines/equipment is maintained properly
- Insure that seats provide proper design features to guard against shocks caused by minor collisions or rough roads that could tend to unseat the involved individual

7.7 Mining Equipment Maintenance Accidents and Causes for Mining Equipment Accidents

There are many types of accidents that occur during mining equipment maintenance activity. Some commonly occurring accidents are contact with hot objects, inhalation of noxious fumes, falling objects, impact with an object or machinery, rolling objects, flying objects, falling on objects, caught, moving objects, overexertion, lifting, and push-pull [19].

Over the years, various studies have been conducted to identify causes for the occurrence of mining equipment accidents. One such study performed by the United States Bureau of Mines (now NIOSH) has identified the following seven causes for the occurrence of mining equipment accidents [12, 20]:

- Poor original design or redesign
- Poor ingress/egress design
- Poor control-display layout

- Restricted visibility
- Unguarded moving parts or components
- Hot surfaces/exposed wiring
- Exposed sharp surfaces/pinch points

7.8 Mining Equipment Safety Analysis Methods

There are many methods in the published literature that can be used to perform safety analysis of equipment used in mines. Five of these methods considered most useful to perform mining equipment safety analysis are consequence analysis, management oversight and risk tree (MORT) analysis, binary matrices, failure modes and effect analysis (FMEA), and preliminary hazards analysis (PHA) [21]. The first three methods are presented below and the remaining two methods are described in Chapter 4.

7.8.1 Consequence Analysis

Consequence analysis is concerned with determining the impact of the occurrence of an undesired event on items such as adjacent property, environment, or people. Some examples of such events are explosions, the release of toxic materials, projection of debris, or fire. The primary consequences of concern in the area of mining include items such as fatalities, injuries, and losses due to property/equipment damage and operational downtime.

Needless to say, this method usually serves as one of the intermediate steps of safety analysis, as the consequences of an accident are normally determined initially using methods such as PHA or FMEA. Additional information on the application of consequence analysis in the mining sector is available in [21].

7.8.2 Management Oversight and Risk Tree (MORT) Analysis

This method first appeared in 1973 and is an effective safety assessment approach that can basically be applied to any safety program in the area of mining [22]. The method focuses on administrative/programmatic control of hazardous conditions and is specifically designed to identify, evaluate, and prevent safety-related omissions, oversights, and errors by workers and management that can result in accidents.

The following main steps are used to perform MORT analysis [12, 22]:

- **Step 1.** Obtain adequate working knowledge of the system under study
- **Step 2.** Select the accident to be analyzed

- **Step 3.** Identify all possible hazardous energy flows and barriers related to the accident sequence
- **Step 4.** Document necessary information in the standard MORT-type analytical tree format
- **Step 5.** Determine factors that cause initial unwanted energy flow
- **Step 6.** Document all the safety program elements that are considered to be less than adequate in regard to the unwanted energy flow
- **Step 7.** Continue conducting analysis of the safety program elements in regard to the rest of the unwanted energy flows (if any)
- **Step 8.** Determine all the management system factors associated with the potential accident
- **Step 9.** Evaluate the accomplished analysis for all safety program elements that could be helpful in reducing the likelihood of the potential accident occurrence

Some of the main benefits of the MORT analysis are as follows [12, 22]:

- Useful to examine human, management, and hardware aspects of an industrial system because they collectively cause accidents
- Comprehensive and effective tool that attempts to evaluate all aspects of safety in any work activity
- Results of MORT analysis can suggest necessary improvements to an ongoing safety program that could be quite useful to save lives, reduce injuries, and reduce property damage

In contrast, some of the main drawbacks of the MORT analysis approach are as follows [12, 22]:

- A time consuming method
- Creates a great amount of complex detail
- Emphasizes management's responsibility for providing a safe work environment

The application of this method to a mining system is demonstrated in [21].

7.8.3 Binary Matrices

This is a useful, logical, and qualitative method to identify system interactions [23]. The approach can be used during the system-description stage of safety analysis or as a final checkpoint in a FMEA or PHA, insuring that each and every important dependency associated with the system under consideration has been considered in the analysis effectively.

The binary matrix is the specific tool used in binary matrices and it contains information on the relationships between the system elements. The main objective of this matrix is to identify the one-on-one dependencies that exist between all elements of a system under consideration. All in all, the matrix serves as a useful tool for "reminding" the involved analyst that failures in one part of a given system may effect the normal operation of other subsystems in totally distinct areas. The application of the binary matrices approach in the area of mining is demonstrated in [21].

7.9 Hazardous Area Signaling and Ranging Device (HASARD) Proximity Warning System

Over the years, a large number of mining workers working near machinery and powered haulage have been killed [12, 24]. A subsequent study of these fatalities revealed that occasionally workers become preoccupied with operating their own equipment and fail to realize when they stray into or are subjected to rather hazardous conditions. This occurs despite the fact that in most of these cases the workers were generally aware of the surrounding dangers.

In order to overcome such problems NIOSH developed an active proximity warning system called HASARD [24]. This system has repeatedly proven to be an effective tool for warning mine workers when they approach hazardous areas/zones around heavy equipment and other hazardous work areas/zones.

The system is basically composed of two subsystems: transmitter and receiver [24]. The transmitter generates a 60 Hz magnetic field with the aid of one or more wire loop antennas. In turn, each antenna is adjusted to establish a magnetic field pattern for each hazardous zone/area, as the need arises.

The receiver is a magnetic field meter and is worn by the mining workers. It compares the received signal with preset levels, which are calibrated to identify danger levels. Furthermore, its outputs can be made to stop ongoing machine/equipment operations and can include audible, visual, and vibratory indicators. Additional information on HASARD proximity warning system is available in [24].

7.10 Problems

1. Write an essay on mining equipment safety.
2. List at least five mining equipment safety-related facts and figures.
3. Discuss equipment fire-related mining accidents.
4. What are the best strategies for reducing mining equipment fires?
5. Discuss mining ascending elevator accidents.
6. List at least ten human factor-related tips concerned with safer mining equipment.
7. Discuss mining equipment maintenance-related accidents.
8. What are the main causes for the occurrence of mining equipment accidents?
9. Discuss consequence analysis.
10. Describe the HASARD proximity warning system.

References

1. Mine Safety and Health Administration (MSHA). US Department of Labor. Washington, D.C. Available from: URL: http://www.msha.gov/.
2. Burgess-Limerick R, Steiner L. Preventing Injuries: Analysis of Injuries Highlights High Priority Hazards Associated with Underground Coal Mining Equipment. Am Longwall Mag 2006; 8:19–20.
3. MSHA Data for 1978–1988. Mine Safety and Health Administration (MSHA). US Department of Labor. Washington, D.C.
4. Cawley JC. Electrical Accidents in the Mining Industry, 1990–1999. IEEE Trans Ind Appl 2003; 39(6):1570–1576.
5. De Rosa M. Equipment Fires Cause Injuries: Recent NIOSH Study Reveals Trends for Equipment Fires at U.S. Coal Mines. Coal Age 2004; 10; 28–31.
6. Rethi LL, Barett EA. A Summary of Injury Data for Independent Contractor Employees in the Mining Industry from 1983–1990. Report No. USBMIC 9344. US Bureau of Mines. Washington, D.C.; 1983.
7. Sanders MS, Shaw BE. Research to Determine the Contribution of System Factors in the Occurrence of Underground Injury Accidents. Report No. USBM OFR 26-89. US Bureau of Mines (USBM). Washington, D.C.; 1988.
8. Unger RL. Tips for Safer Mining Equipment. US Department of Energy's Mining Health and Safety Update 1996; 1(2):14–15.
9. What Causes Equipment Accidents?. National Institute for Occupational Safety and Health (NIOSH), 2008. Available from: URL: http://www.cdc.gov/niosh/mining/topics/machinesafety/equipmentdsgn/equipmentaccident.
10. Kecojevic V, Komljenovic D, Groves W, Rodomsky M. An Analysis of Equipment-Related Fatal Accidents in U.S. Mining Operations: 1995–2005. Safety Science 2007; 45:864–874.
11. Sacks HK, Cawley JC, Homce GT, Yenchek MR. Feasibility Study to Reduce Injuries and Fatalities Caused by Contact of Cranes, Drill Rigs, and Haul Trucks with High-Tension Lines. IEEE Transactions on Industry Applications 2001; 37(3):914–919.
12. Dhillon BS. Mining Equipment Reliability, Maintainability, and Safety. London: Springer Inc.; 2008.
13. Barkand TD. Ascending Elevator Accidents: Give the Miner a Brake. IEEE Transactions on Industry Applications 1992; 28(3):720–729.
14. Nederbragl JA. Rope Brake: As Precaution Against Overspeed. Electrical World 1989;7: 6–7.
15. Barkand TD. Helfrich, Application of Dynamic Braking, to Mine Hoisting Systems. IEEE Transactions on Industry Applications. 1988; 24(5):507–514.
16. Barkand TD. Safe Electrical Design of Mine Elevator Control Systems. IEEE Transactions on Industry Applications 1997; 33(2):312–318.
17. Zeng W. Exploration for Human Factors in the Design of Coal-Mine Safety and Rescue Devices. Proceedings of the 7th International Conference on Computer-Aided Industrial Design 2006:412–416.
18. Unger RL. Tips for Safer Mining Equipment. Holmes Safety Association Bulletin. October; 1996.
19. Unger RL, Conway K. Impact of Maintainability Design on Injury Rates and Maintenance Costs for Underground Mining Equipment in Improving Safety at Small Underground Mines. compiled by RH Peters. Report No. 18-94, US Bureau of Mines. Department of the Interior, Washington, D.C.; 1994.
20. Sanders MS, Shaw BE. Research to Determine the Contribution of System Factors in the Occurrence of Underground Injury Accidents. Report No. USBM OFR 26-89. US Bureau of Mines. Washington D.C.; 1988.
21. Daling PM, Geffen CA. User's Manual of Safety Assessment Methods for Mine Safety Officials. Report No. BuMines OFR 195-(2)-83 US Bureau of Mines. Department of the Interior. Washington D.C.; 1983.

22. Johnson WG. The Management Oversight and Risk Tree-MORT. Report No. SAN 821-1. Washington, D.C.: US Atomic Energy Commission; 1973.
23. Cybulskis P. *et al.* Review of Systems Interaction Methodologies. Report No. NUREG/CR-1896. Columbus, Ohio: Battelle Columbus Laboratories; 1981.
24. Hazardous Area Signalling and Ranging Device (HASARD). Pittsburgh Research Laboratory. National Institute for Occupational Safety and Health (NIOSH). Atlanta, Georgia.

Chapter 8
Electrical Accidents in Mines and Programmable Electronic Mining System Safety

8.1 Introduction

Electricity plays an important role in the mining industries as much of the equipment used in this industrial sector is electrically powered. Over the years many electrical accidents have occurred in the mining industry. According to a study based on the United States Mine Safety and Health Administration (MSHA) coal, metal, and non-metal operator- and contractor-reported data, there were 1926 mine electrical accidents during 1990–1999 [1]. These accidents resulted in 75 deaths and a large number of injuries.

Due to competition and other factors, mining industries have begun using programmable electronics (PE) in areas such as mine monitoring systems, automated haulage, long wall mining systems, and mine processing equipment. As in other industrial sectors, the application of PE technology in the mining industries has created various challenging issues for system design, verification, operation, maintenance, and assurance of functional safety [2]. More specifically, although there are many benefits to using PE, it adds a level of complexity that may compromise occupational safety if not handled with care [3–5]. This chapter presents various important aspects of electrical accidents in mines and programmable electronic mining system safety.

8.2 Fatal Electrical Accidents in Comparison to Other Fatal Mine Accidents and in Different Mining Areas

Between 1990–1999 there were 1926 electrical accidents in United States mines, of which 75, or 3.89%, were fatal [1].

The percentages of fatal accidents in other mine classifications (in parentheses) during the same time period were 10.69% (falling/rolling/sliding rock or material of any kind), 7.92% (explosives and breaking agents), 2.39% (fall of face/rib/side/high wall), 2.03% (exploding vessels under pressure), 1.85% (ignition/explosion of gas or dust), 1.42% (powered haulage), 0.54% (fall of roof or back), 0.53% (non-electrical or explosion-associated fire), 0.44% (machinery), 0.37% (inundation), 0.15% (slip or fall of person), 0.13% (hoisting), 0.05% (hand tools), and 0.01% (handling material) [1].

During this period there were 75 fatal electrical accidents. The distribution of these fatalities during this period among different mining sectors were as follows: coal (Bituminous), 33, Limestone (crushed and broken), 8, Sand and Gravel, 7, Copper Ore, 6, Cement, 1, Gold (iode and placer), 3, Clay (common), 2, Granite (crushed and broken), 2, Phosphate rock, 2, Traprock (crushed and broken), 2, Potash, 2, Lead and/or Zinc Ore, 1, Stone(crushed and broken), 2, Iron ore, 0, Alumina (Mill), 0, Trona, 1, and Coal (anthracite), 3 [1].

8.3 Nature of Injury from Electrical Accidents in Mines and Job Titles of Victims of Mining Electrical Accidents

Although electrical accidents cause various types of injuries, between 1990–1999 around 65% of the electrical accidents in the United States mines caused burn injuries and only 7% of these accidents caused fatalities [1]. On the other hand, 24% of the electrical accidents caused electric shock (electrocution) injuries but 93% of electrical fatalities.

A breakdown of the different types of injuries resulting from mine electrical accidents for this period are as follows [1, 5]:

- Radiation effects (burn from electrical arc-not contact), 770
- Electrical shocks (electrocution), 447
- Burns (electrical), 164
- Burns or scaldings (heat-not radiation), 128
- Burns (chemical), 126
- Multiple injuries, 45
- Sprains/strains/ruptured discs/whiplash/torn knee cartilage, 32
- Cuts/lacerations/punctures/infections, 28
- Asphyxia/strangulation/drowning/smoke inhalation/suffocation, 17
- Fractures, chips, 16
- Contusions, bruises, 11
- Scratches, abrasions (superficial wounds), 9
- NEC (Not Elsewhere Classified), 9
- Amputation or enucleation: 7 injuries
- Poisonings, systemic, 6
- Dust or other particles in eyes, 3
- Concussions (brain, cerebral), 2
- Dislocations, 2

- Hearing loss or impairment (industrial), 1
- Hernia ruptures, 1
- Radiation effects, not elsewhere classified, 1
- Radiation effects (sunburn), 1

Electrical accidents in mines do not just injure the persons involved in electrical work, but many others as well. For example, between 1990–1999 mining electrical accidents in the United States injured persons performing a variety of tasks. Some of the regular job titles of these individuals were electrician, roof bolter/rock bolter, truck driver (surface), welder (surface), oiler/greaser (surface), bulldozer operator, high lift/front end loader operator, shuttle car operator, mine manager/mine foreman, repairman/mechanic (surface), maintenance foreman, muck machine operator, belt/conveyor man (off section), laborer/utility man, and mining engineer [1, 6].

8.4 Activities Performed During the Occurrence of Mining Electrical Accidents and Equipment Involved in Mine Electrical Accidents

Experience indicates that there were a wide range of activities being performed when mining electrical accidents have occurred. For example, between 1990–1999 some of the work activities reported as being performed when mining electrical accidents occurred in the United States were as follows [1, 5]:

- Maintenance/repair (electrical)
- Maintenance/repair (machinery-not electrical)
- Moving of power cable (includes reeling)
- Welding and cutting
- Operation of locomotive (air trimmer)
- Handling of supplies/materials (not timber)-loading/unloading
- Observation of operations
- Inspection of equipment (not maintenance/repair)
- Moving of equipment (fans/pumps, not operating machinery)
- Operation of mill equipment
- Getting on/off equipment, machines, etc.
- Escaping a hazard
- Rerail equipment (includes replacing trolley poles)

It should be noted that during this period electrical maintenance/repair work was the most hazardous activity. It accounted for around 50% of all electrical accident-related injuries and about 48% of all electrical accident-related fatalities [1]. Furthermore, during the same time period maintenance/repair (machinery-not electrical) was the second most hazardous work activity. It accounted for roughly 13% of all electrical accident-related injuries and approximately 7% of all accident-related fatalities.

Various types of equipment have also been involved in mine electrical accidents. As an example some of the types of equipment involved in mining electrical accidents in the United States between 1990–1999 were as follows [1, 6]:

- Welding machines
- Cranes, boom hoists, cherry pickers, etc.
- Front-end loaders, payloaders, highlifts, etc.
- Rock or roof bolting machines
- Pumps
- Fans
- Tunnel borers
- Man cars, personnel carriers, porta buses, etc.
- Machines, NEC (Not Elsewhere Classified)
- Conveyors, belt feeders, etc.
- Ore haulage trucks
- Crushers
- Shovels or draglines (mining and striping)
- Bulldozers
- Drills (carriage-mounted), on track/rail/rubber tired

8.5 Measures for Mitigating Mine Electrical Shock Injuries in General and in Maintenance Work

Electrical shock and electrocution are responsible for a significant percentage of electrical injuries in mines. For example, during 1990–1999 electrical shock and electrocution caused around 23% of all mine electrical injuries in the United States [1]. Furthermore, electrical shock and electrocution injuries resulted in 93% of all mine electrical fatalities.

According to various investigations electrical shock and electrocution injuries are frequently the result of working with live or in proximity to unguarded live electrical conductors. Some of the possible measures for mitigating electrical shock/electrocution injuries are as follows [1, 6]:

- Maintain appropriate clearance when working close to overhead electric power lines
- Use ground-fault circuit interrupters (GFCIs) as much as possible
- Use power line proximity and/or contact warning systems
- Use insulating load link devices

Maintenance is an important activity in the area of mining. Maintenance and repair work caused around 60% of all mine electrical accidents and about 53% of all mine electrical fatalities [1]. Some of the possible measures to mitigate maintenance work electrical-related injuries are as follows [1, 5, 6]:

- Use GFCIs as much as possible
- Use "dead-front" type equipment to isolate maintenance personnel from electrical hazards during the troubleshooting process
- Require each electrical enclosure to have a single disconnect mechanism or interlock that de-energizes all circuits within an enclosure

Often mine electrical maintenance workers use meters for troubleshooting live electrical circuits. A study of electrical accident data indicates that there is a definite need for a better method that accurately verifies meter functions and capabilities in the field, and to avoid using incorrect voltage rating meters or meters set to measure the incorrect function [1]. Some useful guidelines for improving the safe application of electrical meters during live troubleshooting processes are as follows [1, 6]:

- Clearly mark or color code meters with their maximum current/voltage ratings
- Use test leads that are rated for the maximum current/voltage of the associated meter
- Use only those safety test leads that have minimal tip exposure to preclude accidental contact with adjacent circuits
- Use only those meters that auto-range up to their maximum current/voltage to prevent range selection-related difficulties

8.6 Programmable Electronic-related Accidents in Mines

The safety of programmable electronics (PE) based systems has become an important issue with their increased use in the mining sector. During 1995–2001, eleven PE-related incidents occurred in United States mines. Four of these incidents resulted in fatalities [2, 3, 8]. During the same time period, 71 PE-related incidents occurred in underground coal mines in New South Wales, Australia [2, 3].

A study of these two sets of data indicated that the most of incidents involved sudden start-ups or movements of PE-based mining systems. In 1991, an MSHA study of all long wall installations with respect to PE reported that 35% experienced sudden movements basically due to four problems: water ingress, software programming errors, sticking or defective solenoid valves, and operator error [2, 3, 9].

In addition, study of these data sets revealed that there were four major factors that contributed to PE-based mishaps: improper operation, software, water ingress, and solenoid valves [3, 9, 10]. Furthermore, the study revealed solenoid valve problems as the main cause for the PE-based mishaps.

Over the years MSHA has taken various measures to reduce PE-based mishaps including proposing a safety framework largely based on the IEC 61508 safety life cycle document [11] for complex PE mining systems.

8.7 Methods for Conducting Programmable Electronic Mining System Hazard and Risk Analysis

There are many methods that can be used to conduct programmable electronic mining system hazard and risk analyses. The methods considered most useful are preliminary hazard analysis (PHA), hazard and operability studies (HAZOP), failure mode and effect analysis (FMEA), interface analysis, fault tree analysis (FTA), action error analysis (AEA), event tree analysis (ETA), operating and support analysis (OASA), potential or predictive human error analysis, and sequentially timed events plot (STEP) investigation systems [3, 12]. Interface analysis and the PHA, HAZOP, FMEA, and FTA methods are described in Chapter 4 and the remaining ones are presented below in separate subsections [3, 13].

8.7.1 Operating and Support Analysis

This is a quite useful method and is concerned with identifying potential hazards during operation and maintenance, finding the associated root causes, determining the acceptable level of risk, and recommending effective measures for reducing risk. OASA is conducted by a group of people who are quite familiar with the system's operation and interaction with all involved persons.

Some of the factors considered during OASA are shown in Figure 8.1 [13, 14].

The major benefit of OASA is that it is an effective tool to provide hazard identification in the context of total system operation. Similarly, its main drawback is that it requires a rather high degree of expertise concerning the system under consideration. Additional information on this method is available in [13, 14].

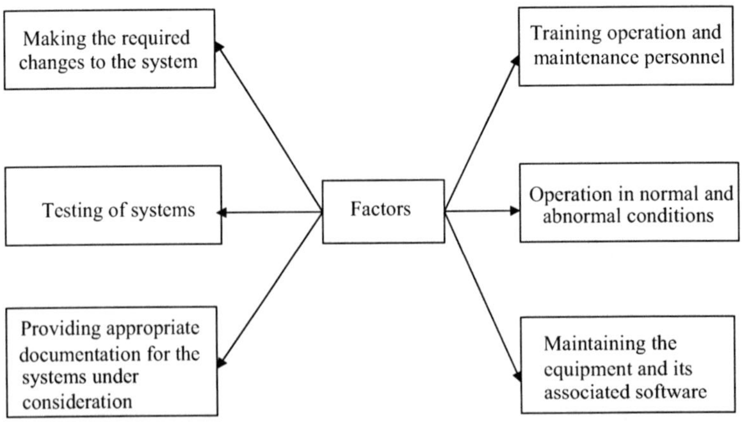

Figure 8.1 Some of the factors considered during operating and support analysis

8.7.2 Action Error Analysis

This method is concerned with identifying operator errors and their consequences. The approach focuses on the interactions between a system under consideration and all involved humans during testing, operation, and maintenance phases. For operation and maintenance-related tasks action error analysis is carried out by following the three basic steps [14]:

- **Step 1.** Highlight operator tasks
- **Step 2.** Describe each operator task associated subtasks and actions
- **Step 3.** Identify all possible operator errors and their consequences for each action

The types of errors considered for each action in Step 3 are shown in Figure 8.2 [3, 14].

The principal advantage of this method is that it is extremely useful for semi-automated or automated processes having operator interfaces. In contrast, its main drawback is that it requires a high degree of expertise concerning the system under consideration. Additional information on the method is available in [14].

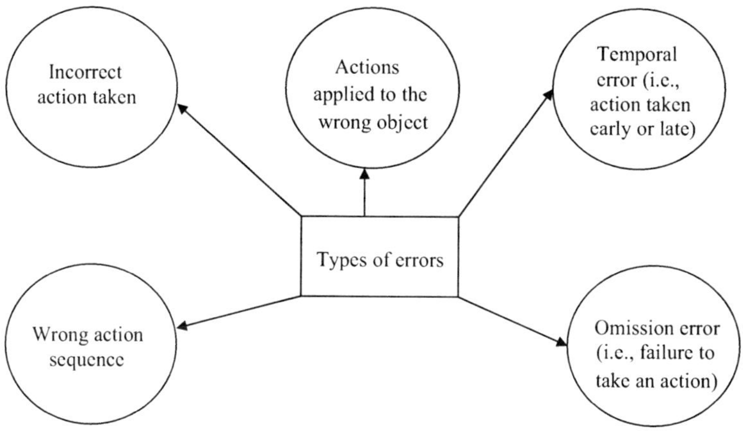

Figure 8.2 Types of errors considered for each action

8.7.3 Potential or Predictive Human Error Analysis

This method is similar to the HAZOP method and is a team based approach. It focuses on human tasks and their associated error potential and classifies human error causes under five basic categories. These categories along with their corresponding effects are presented in Table 8.1 [3, 15].

Table 8.1 Human error cause categories and their corresponding effects

Category No.	Human error cause	Effect
1	Stress	Increases the likelihood of error
2	Complexity	Increases the likelihood of error
3	Fatigue	Increases the likelihood of error
4	Environment	Adverse environments increase the likelihood of error
5	Training	Better training decreases the likelihood of error

Predictive human error analysis is performed in two steps as follows [13]:

- **Step 1.** Highlight all key human tasks
- **Step 2.** For each task, use guide words such as incorrect action timing, wrong action, incomplete action, action applied to wrong interface object, and incorrect action sequence

The major advantage of this approach is that it can highlight a large proportion of potential errors. In contrast, it has two major drawbacks [13]. These are that its effectiveness depends on the expertise and effort of team members, and that it can be a quite time-consuming method when there are many tasks and actions.

8.7.4 Event Tree Analysis

This is a "bottom up" method that identifies the possible outcomes when the occurrence probability of the initiating event is known. The approach has proven to be a useful tool for performing analyses of facilities having engineered accident-mitigating characteristics, for identifying the sequence of events that follow the initiating event, and for generating given sequences. Usually, it is assumed that each sequence event is either a success or a failure.

It should be noted that because of the inductive nature of the method, the basic question addressed is, "What happens if …?" In comparison to the FMEA approach described in Chapter 4, ETA is generally used to perform analysis of more complex systems. Two principal advantages of this method are that it is a powerful tool for single events with multiple outcomes and for rather high risks not amenable to simpler analysis approaches [3, 13]. Similarly, two main disadvantages of the method are that it is a time-consuming approach and that the occurrence probabilities of events may be difficult to estimate [3, 13]. Additional information on this method is available in [16–18].

8.7.5 Sequentially Timed Events Plot Investigation System

This is an analytical approach that graphically depicts sequentially timed events.

All events are expressed with formatted "building blocks" composed of an "actor and action" [3, 13]. The method is considered quite useful to discover and analyze problems and to assess mitigation options, in addition to performing analysis of the types and sequences of events that can lead to an incident.

Two major benefits of this method are that it can be used to define and systematically analyze complex systems or processes, and that it facilitates focus-group analysis. The major drawback of the method is that it is normally perceived as a complex technique that is rather costly to implement. Additional information on the method is available in [13].

8.8 Lessons Learned in Addressing Programmable Electronic Mining System Safety-related Issues

Over the years, many lessons have been learned in addressing programmable electronic mining system safety-related issues. Some of these lessons are as follows [19]:

- Highlight and clearly understand all related perceptions and issues
- Involve all important elements of the industrial sector as early as possible and on a continuous basis
- Develop appropriate definitions, terminology, and concepts as early as possible
- Clearly separate all of the associated concerns
- Divide the problem into a number of manageable elements
- Hold necessary industry workshops
- Use appropriate scenarios for conveying information

Additional information on this topic is available in [19].

8.9 Problems

1. Write an essay on electrical accidents in mines.
2. Discuss the nature of injury from electrical accidents in mines.
3. List at least ten job titles of victims of mining electrical accidents.
4. List at least ten activities that were being performed at the occurrence of mining electrical accidents.
5. List at least fifteen types of equipment involved in mine electrical accidents.
6. Discuss measures for mitigating mine electrical shock injuries in general and in maintenance work.
7. Discuss programmable electronic-related accidents in mines.
8. List at least nine methods that can be used to perform programmable electronic mining system hazard and risk analysis.
9. Discuss the following two methods:
   - OASA
   - AEA
10. Compare ETA with potential or predictive human error analysis.

References

1. Cawley JC. Electrical Accidents in the Mining Industry, 1990–1999. IEEE Transactions on Industry Applications 2003;39(6):1570–1577.
2. Sammarco JJ. Addressing the Safety of Programmable Electronic Mining Systems: Lessons Learned. Proceedings of the 37th IEEE Industry Applications Society Meeting 2003:692–698.
3. Dhillon BS. Mining Equipment Reliability, Maintainability, and Safety. London: Springer, Inc.; 2008.
4. Sammarco JJ, Kohler JL, Novak T, Morely LA. Safety Issues and the Use of Software-Controlled Equipment in the Mining Industry. Proceedings of the 32nd IEEE Industry Applications Society Meeting 1997:496–502.
5. Sammarco JJ. Programmable Electronic Mining Systems: Best Practice Recommendations (in nine parts). Report No. IC 9480 (Part 6: 5.1 System Safety Guidance). National Institute for Occupational Safety and Health (NIOSH). US Department of Health and Human Services. Washington DC; 2005.
6. Mary CS, Floyd L, Eastwood K, Liggett D. How Can We Better Learn From Electrical Accidents. IEEE Ind Applcat Mag 2000;6(May/June):16–23.
7. Rossignol M, Pineault M. Classification of Fatal Occupational Electrocutions. Can J Public Health 1994;85(5):322–325.
8. Fatal Alert Bulletins, Fatal Grams and Fatal Investigation Reports. Mine Safety and Health Administration (MSHA). Washington, D.C.(May 2001). Available from: URL: www.msha.gov/fatals/fab.htm.
9. Dransite GD. Ghosting of Electro-hydraulic Long Wall Shield Advanced Systems. Proceedings of the 11th West Virginia University International Electrotechnology Conference 1992:77–78.
10. Waudby JF. Underground Coal Mining Remote Control of Mining Equipment: Known Incidents of Unplanned Operation in New South Wales (NSW) Underground Coal Mines. Dept. of Mineral Resources, NSW Dept. of Primary Industries, Maitland, NSW, Australia; 2001.
11. IEC 61508. Parts 1–7. Functional Safety of Electrical/Electronic/Programmable Electronic Safety-Related Systems. International Electrotechnical Commission (IEC). Geneva, Switzerland 1998.
12. Defense Standard 00-58 (Parts I and II). HAZOP Studies on Systems Containing Programmable Electronics. Directorate of Standardization. UK Ministry of Defence. Glasgow, UK 1998.
13. Sammarco JJ. Programmable Electronic Mining Systems: Best Practice Recommendations (in nine parts). Report No. IC9480 (Part 6: 5.1 System Safety Guidance). National Institute for Occupational Safety and Health (NIOSH). US Department of Health and Human Services. Washington, D.C., 2005. Available from NIOSH: Publications Dissemination, 4676 Columbia Parkway, Cincinnati, Ohio 45226, USA.
14. Harm-Ringdahl L. Safety Analysis: Principle and Practice. London: Elsevier, 1993.
15. Guidelines for Preventing Human Error in Process Safety, Center for Chemical Process Safety, American Institute of Chemical Engineers, New York, 1994.
16. CAN/CSA-Q6340-91, Risk Analysis Requirements and Guidelines, Prepared by the Canadian Standards Association, Toronto, 1991. Available from the Canadian Standards Association (CSA), 178 Rexdale Boulevard, Rexdale, Ontario, Canada.
17. Dhillon BS. Engineering Safety: Fundamentals, Techniques, and Applications. River Edge, New Jersey: World Scientific Publishing; 2003.
18. Cox, SJ, Tait NRS. Reliability, Safety, and Risk Management. Oxford, U.K: Butterworth-Heinmann; 1991.
19. Sammarco JJ. Addressing the Safety of Programmable Electronic Mining Systems: Lessons Learned. Proceedings of the 37th IEEE Industry Applications Society Meeting 2003: 692–698.

Chapter 9
Gas-related, Fire, and Blasting Accidents in Mines and Methods for Determining Mine Atmosphere Status

9.1 Introduction

Coal bed gas has been considered a major mine hazard since the first documented coal mine gas explosion in 1810 in the United States [1]. It has considerably affected safety and productivity in underground coal mines throughout the world. Fires are another major safety problem in mines. For example, there were over 7,700 fires during the period between 1947–2006 in underground hard coal mines in Poland alone [2].

Blasting is an important and hazardous element of mining and many serious injuries and fatalities result from improper judgment or practice during the blasting process. During 1990–1999, approximately 22.3 billion kilograms of explosives were used by the mining, quarrying, construction, and other industrial sectors in the United States [3].

Samples of mine atmosphere are usually taken during normal operations to establish a reliable baseline. On the basis of these samples, mine operators can effectively determine changes in mine atmosphere. Furthermore, in the event of a fire-related emergency these samples become very useful in fire fighting operations, rescuing operations, etc. This chapter presents various important aspects of gas-related, fire, and blasting accidents in mines and two important methods for determining mine atmosphere status.

9.2 Origin and Mechanism of Coal Mine Outbursts and Their Prediction and Prevention

It is commonly accepted that by virtue of their chemical and physical characteristics, rock aggregates in the earth's crust are composed of a network of structures that include pores, fractures, and micro cracks filled with liquid and gaseous substances. Coal is one example of these rock aggregates that has accumulated gaseous substances such as methane, carbon dioxide, and/or nitrogen in these structures during the coalification process.

Some of the direct or indirect causes identified for outbursts in underground coal mines are as follows [1]:

- The internal energy of the gas contained in the coal bed
- Gas pressure and quantity
- Rock pressure and strength
- Micro seismic activity propagated by reactivation of faults and explosives
- Maceral composition of the coal

Knowledge accumulated through investigations to determine the origin of coal bed gas and mechanisms of coal mine outbursts has played an instrumental role in enhancing outburst and explosion prediction and prevention methods. Prediction includes activities such as monitoring of gas emissions, monitoring of microseismic acoustic emissions, and monitoring of acoustic signal spectrum characteristics with respect to structures in the rock mass [1].

Some of the methods that can be used to prevent the occurrence of coal-mine outbursts are as follows [1]:

- **Utilizing predicting methods.** These methods can be used to control and prevent underground coal mine outbursts. For example, by taking gas content and pressure measurements from drill holes on the surface and subsurface, one can determine the threshold conditions for outburst occurrence.
- **Modifying mining methods and machinery.** This approach is concerned with the modification of the existing methods and machinery to take into consideration rock-mass stresses that could trigger coal bed gas outbursts from an advancing mine face.
- **Adopting different methods for gas ventilation or drainage.** Two examples of such methods are removing gas through vertical wells in advance of mining, and drilling vertical gob wells into the cave area behind the long wall panel.

Finally, it should be noted that the application of the above methods can be useful to improve mine safety, efficiency of mine operations, and mine economics.

9.3 Underground Fires in Hard Coal Mines and Out-of-Control Coal Fires

Hard coal is widely used in generating electricity and steel production throughout the world. In 2001, the top ten hard coal-producing countries were China, USA, India, Australia, South Africa, Russia, Poland, Indonesia, the Ukraine, and Kazakhstan [4]. These countries produced 1294, 945, 312.5, 257, 224.5, 168, 104, 92.5, 82, and 73 million tons of hard coal in 2001, respectively.

Table 9.1 Decade breakdowns of underground fires in Polish hard coal mines for the period 1947–2006

Decade No.	Time period (Years)	No. of fires
1	1947–1956	4427
2	1957–1966	2627
3	1967–1976	283
4	1977–1986	230
5	1987–1996	136
6	1997–2006	54

Coal fires are a major safety problem around the globe. Some of the mine-related activities responsible for igniting fires in most hard coal mines are as follows [4]:

- Welding
- Cutting
- Electrical Work
- Explosives

Needless to say, work activities such as these and others have resulted in thousands of underground fires in hard coal mines around the world. For example, in Poland alone there were a total of 7757 underground fires during the period 1947–2006 in hard coal mines [2]. Their decade breakdowns are presented in Table 9.1 [2].

It is interesting to note from Table 9.1 that the number of fires has decreased quite dramatically over the stated time period. However, in spite of many safety-related achievements in combating underground fires in hard coal mines, many still occur.

Out-of-control coal fires around the globe are an environmental catastrophe characterized by factors such as the emission of noxious gases, condensation by-products, and particulate matter. Some of the oldest, largest, and environmentally catastrophic out-of-control coal fires in the world are in China, USA, and India [4]. Fires in each of these three countries are discussed below [4].

- **China.** China leads the world in coal production, accounting for one third of the global output. A large number of out-of-control coal fires are in the northern part of China, particularly in Xinjiang and Ningxia provinces.
  Underground coal fires in the Liu Huangou coalfield in Xinjiang province have been burning over the past 20 years and the province's capital city, Urumqi, is considered one of the most polluted cities in the world. In the Rujigou coalfield of Ningxia province, underground coal fires are clearly considered responsible for land subsidence as well as for the release of hydrogen sulphide into the atmosphere. All in all, atmospheric pollution throughout China is considered among the highest in the world, and is primarily from coal combustion [4].
- **USA.** The United States is the second largest coal producer in the world and has many out-of-control underground coal mine fires, particularly in the state of Pennsylvania [4]. Although Pennsylvania ranks fourth in the United States coal

production, after the states of Wyoming, West Virginia, and Kentucky over one third of the United States abandoned mine-associated problems occur in this state and coal fires are considered among the worst such problems.
Coal fires in Pennsylvania have been recorded since 1772 and currently there are 140 underground coal mine fires in the state [4]. In particular, the Percy coal mine fire in Youngstown, Pennsylvania has been burning for more than 30 years.

- **India.** India is the third largest coal producing country in the world and has many out-of-control coal mine fires, particularly in the state of Bihar. The first coal fire in Jharia coal field at Bhowrah, Bihar broke out in 1916. Currently, there are around 70 fires burning in the Jharia coal field. Most fires in this coal field were due to the spontaneous combustion of coal subsequent to open cast and deep mining, and it is estimated that around 37 million tons of coal has been lost due to the Jharia coal field fires [4].

9.4 Use of Inert Gases in Mine Fires

Inert gases have been used for over 150 years as a means of assisting in the fighting of mine fires [5]. Some of the ways inert gases can be used to fight mine fires are as follows [5]:

- To extinguish fires by reducing the oxygen concentration in the surrounding atmosphere
- To prevent the occurrence of gas explosions in situations when fires are fought directly or when seals are closed or constructed
- To effectively seal fire zones with pressure chambers

It should be noted that in theory any inert gas can be used for generating an extinctive atmosphere in a fire zone area. However, in practice the choice narrows to just three types of gases: carbon dioxide, flue gas from combustion, and nitrogen. Over the years many disadvantages have been noticed in using such extinctive gases. Some of these disadvantages are as follows [5–7]:

- It becomes very difficult to assess the state of the fire from the analysis of gases
- When in contact with incandescent material such gases may act as an oxygen donor
- Under the above conditions, carbon dioxide can generate or produce a large amount of carbon monoxide
- The density differences between the mine atmosphere and inert gases can result in ventilation disturbances. This, in turn, can let the inert gases' mixture and fume enter airways other than the return airways
- If a fire zone is ventilated with oxygen prior to the closing of seals, some amount of gas may flow from the return airways in form of layers along the floor into adjoining inclined roadways, if any exist

Additional information on the use of inert gases in mine fires, is available in [5].

9.5 Blasting Injuries in Surface Mining and Blast Damage Index

Blasting is a complex activity and a hazardous element of surface mining. The hazards associated with surface blasting are basically due to the following four factors [3]:

- Flyrock
- Lack of blast area security
- Premature blast
- Misfire

The term "flyrock" may simply be defined as the rock propelled beyond the blast zone or area by the force of an explosion. Some of the main factors responsible for flyrock are inadequate blast hole layout and loading, poor stemming, anomalies in the geological and rock structure, and insufficient burden. Injuries due to fly rocks are sustained when they travel beyond the blast area or zone and strike people. With regard to the lack of blast area security, some of the injury causing factors are failure to use proper blasting shelters, inadequate guarding of the blast area, and poor communication [3, 8].

During the period between 1978–1998, a total of 45 fatal and 367 non-fatal blasting injuries in surface mines, were reported in the United States [9]. The annual numbers of fatal injuries, shown in parentheses, were 1978 (2), 1979 (2), 1980 (3), 1981 (5), 1982 (3), 1983 (0), 1984 (4), 1985 (0), 1986 (3), 1987 (1), 1988 (0), 1989 (5), 1990 (5), 1991 (1), 1992 (4), 1993 (1), 1994 (3), 1995 (1), 1996 (1), 1997 (1), and 1998 (0).

A study of the occurrence of injuries indicates that the lack of attention to three main factors has the potential to cause injury [3]. These factors are task, personal, and environmental factors, and are depicted in Figure 9.1.

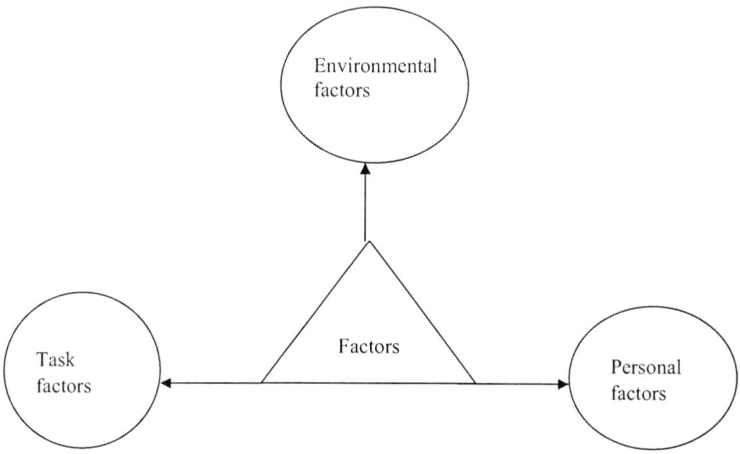

Figure 9.1 Major factors involved in blasting having the potential to cause injury

Task factors include items such as clearing of the blast site, clearing of the blast area, examination of the blast site, sound warning signals, protection in a blasting shelter, supervision of loading, securing the blast area, and supervision of stemming.

Similarly, personal factors include items such as job experience, training, visual perception, loading and blasting in a hurried manner, prior injury history, education, and overwork.

Finally, environmental factors include items such as geological anomalies, uneven ground, misfires, drilling noses, equipment movement, obstruction of visibility, lightning, movement of explosive vehicles, and smoke, dust, and gases.

9.5.1 *Blast Damage Index (BDI)*

The Blast Damage Index (BDI) is concerned with assessing the damage to underground openings in coal mines due to open-cast blasting. The index is analogous to the reciprocal of safety factor and is expressed by [10]

$$BDI = \frac{IS}{DR} = \frac{V\theta v_{\mathrm{cr}}}{C_{\mathrm{q}}\beta_{\mathrm{dt}}} \tag{9.1}$$

where

- BDI is the blast damage index.
- IS is the induced stress.
- DR is the damage resistance.
- V is the vector sum of peak particle velocity, expressed in mm/s.
- θ is the density of rock mass, expressed in kg/m^3.
- v_{cr} is the compressional wave velocity of rock mass, expressed in m/s.
- C_{q} is the site quality constant.
- β_{dt} is the dynamic tensile strength of rock mass, expressed in N/m^2 or Pascal.

The values of *BDI* are interpreted as follows [10]:

- >2: Major damage
- <1: No damage
- $1 \leq BDI \leq 2$: Minor damage

Additional information on this index is available in [10].

9.6 Flammable Gas Explosions and Useful Recommendations for Their Avoidance

Over the years many flammable gas-related explosions have occurred in various types of mines around the world. For example, flammable gas caused 78 accidents

Table 9.2 Flammable gas-related fatalities and injuries in South African coal, gold, and platinum mines for 1988–2005

Type of mine	No. of injuries	No. of fatalities
Coal	52	8
Gold	25	67
Platinum	67	14
Total	144	89

in South African coal, gold, and platinum mines during 1988–2005 [11]. These accidents resulted in 89 fatalities and 144 injuries. Distribution of the occurrences of these fatalities and injuries according to the type of mine are presented in Table 9.2 [11]. Forty four of the fatalities occurred in three explosions in three different mines. Recommendations made to avoid the reoccurrence of explosions such as those that resulted in the fatalities in two of these three mines are presented below [11].

Beatrix Mine

This mine is located in the Theunissen Magisterial District, Province of the Free State, South Africa. The flammable gas explosion occurred on May 15, 2000 and resulted in seven fatalities and considerable damage to mine property. A subsequent investigation of the explosion made recommendations such as listed below to avoid the occurrence of such explosions in the future [11]:

- Hazard identification to be conducted during planning meetings and geological anomalies to be projected on both dip and strike, forewarning development crew members of the potential water and gas traps
- All hazardous areas to be declared by the management, and the use of appropriate explosion protected equipment within these hazardous areas to be made mandatory
- A system for monitoring flammable gas emissions by means of strategically placed detector heads and appropriate data recording apparatus to be installed
- A system for providing sufficient warning to all individuals concerning the presence of flammable and noxious gases in the area and any failures in the ventilating system to be installed.

Mponeng Gold Mine

This South African gold mine is owned by AngloGold. The flammable gas explosion in the mine occurred on July 29, 1999 at a depth of 2737 m below surface and resulted in 19 fatalities and considerable damage to surrounding mine property within a distance of up to approximately 500 m [11]. Some recommendations were made to avoid the occurrence of such explosions in the future as a result of a subsequent investigation into the explosion. These are as follows [11]:

- All rigs used for cover drilling should have capabilities to turn off an inflow of gas/water immediately
- All efforts should be directed to avoiding recirculation of air at all times
- All ventilation systems should be rigorously designed to take into consideration the results of risk assessments
- All auxiliary fans installed beyond the last point of through ventilation should be flameproof, and when flammable gas is intersected, the normal level of airflow is maintained

Additional information on these recommendations is available in [11].

9.7 Methods for Determining the Status of Mine Atmosphere

Samples of mine atmosphere are usually taken on a regular basis during normal operations to establish a reliable baseline. More specifically, on the basis of the results of these samples mine operators can effectively determine any changes in mine atmosphere. In the event of fire-related emergencies, these samples become quite critical to persons involved in the fire fighting effort. Furthermore, these samples provide vital information on the status of the mine atmosphere to rescue persons and others [12, 13].

There are two methods that can be used to obtain mine gas samples: real-time instruments and gas chromatography [12]. The real-time time instruments are specifically designed for providing almost instantaneous results. These instruments vary from hand-held sample tubes to larger and more complex instruments. Two examples of these instruments are infrared sensors and stain-tube chemical sensors.

In the case of gas chromatography, the instruments are usually housed in a transportable laboratory or building. In contrast to real-time instruments, in this method the time period needed for analysis is the time interval between sample acquisition and the ultimate analysis results. Gas chromatography is frequently used to compare results obtained through real-time instruments.

Over the years many methods/equations have been developed to help in establishing mine atmosphere status. Those methods/equations/models can be used to determine items such as listed below [12]:

- Is there a fire?
- What is burning?
- Is the fire continuing to burn subsequent to sealing the affected area or mine?
- Is the sealed atmosphere explosive or will it become explosive at the commencement of ventilation?

Two of the methods concerned with the first two items are presented below.

9.7.1 Carbon Dioxide Index

This index was developed by J.I. Graham in 1917 for monitoring elevated carbon dioxide levels and is expressed by [12, 14].

$$ICO_2 = \frac{[CO_2 - \theta]}{(\gamma - O_2)} \tag{9.2}$$

where

θ = 0.03
γ = 100
CO_2 is carbon dioxide. It is considered a primary fire detection gas. Furthermore, carbon dioxide gas is formed in a number of ways and is a main product of combustion.
O_2 is oxygen. In a normal atmosphere, it comprises slightly less than 21% by volume and its density with respect to dry air is 1.105.

This index is based on gas percentages and it indicates elevated carbon dioxide values in an oxygen-free atmosphere. Under normal condition (i.e., normal atmosphere), where only ambient carbon dioxide is produced, the value of Equation 9.2 describes the proportion of carbon dioxide to total inert gas and should approach zero.

When ignition occurs, the value of Equation 9.2 increases and continues to increase as long as the fuel quantity or temperature keeps increasing. In sealed zones having no combustion, the value of Equation 9.2 will initially increase, but then it will stabilize with time. Additional information on this index is available in [12].

9.7.2 Jones-Trickett Ratio

This ratio was developed by J.E. Jones and J.C. Trikett in 1954 and is used to determine sample reliability [15]. The ratio is based on the assumption that the number of molecules of matter consumed in a fire is proportional to the number of molecules of gas produced or generated. In turn, the number of gas molecules produced is assumed proportional to the gas volume by assuming constant temperatures.

The ratio is expressed by [12, 15]

$$R_{jt} = \frac{CO_2 + \theta_1(CO) - \theta_2(H_2)}{\theta_3(N_2) - O_2} \tag{9.3}$$

where

R_{jt} is the Jones-Trickett ratio.
θ_1 = 0.75
θ_2 = 0.25
θ_3 = 0.265

CO_2 is carbon dioxide.
O_2 is oxygen.
CO is carbon monoxide. This gas is frequently used as fire detection gas.
H_2 is hydrogen. This gas serves as another important element in determining if fire exists.
N_2 is nitrogen. This is the most abundant gas and comprises 79% of the ambient atmosphere.

Equation 9.3 is based on percent gas concentrations and its value varies with the type of fire. The values of R_{jt} are interpreted as follows [12]:

- <0.4: no combustion is present
- $0.4 < R_{jt} < 0.5$: methane is the fuel source
- $0.5 < R_{jt} < 0.9$: the fuel source is oil, coal, urethane foam, insulation, or conveyor belt
- $0.9 < R_{jt} < 1.6$: wood is the fuel source

It should be noted that values of R_{jt} greater than 1.6 usually occur only under controlled laboratory conditions. This implies that when values of R_{jt} greater than 1.6 are encountered during a mine fire, the sampling device or gas chromatograph should be examined with care for mathematical or calibration errors. Additional information on this ratio is available in [12, 15].

9.8 Problems

1. Discuss the origins and mechanisms of coal mine outbursts.
2. Discuss at least three approaches that can be used to prevent the occurrence of coal mine outbursts.
3. List world's top ten hard coal-producing countries.
4. What are the mine-related activities responsible for igniting fires in most hard coal mines?
5. What are the countries that have some of the oldest, largest, and environmentally catastrophic out-of-control coal fires? Discuss such fires in two of these countries.
6. What are the ways in which inert gases can be used to fight mine fires?
7. What are the disadvantages of using extinctive gases such as carbon dioxide, flue gas from combustion, and nitrogen?
8. What are the factors that cause hazards in surface blasting?
9. Define blast damage index and list its three interpretations.
10. Define the following two items:

    - Jones-Trickett ratio
    - Carbon dioxide index

References

1. Flores RM. Coalbed Methane: From Hazard to Resource. International Journal of Coal Geology 1998; 35:3–26.
2. Wachowicz J. Analysis of Underground Fires in Polish Hard Coal Mines. Journal of China University of Mining and Technology 2008; 18(3):332–336.
3. Bajpayee TS, Rehak TR, Mowrey GL, Ingram DK. Blasting Injuries in Surface Mining with Emphasis on Flyrock and Blast Area Security. Journal of Safety Research 2004; 35:47–57.
4. Stracher GB, Taylor TP. Coal Fires Burning Out of Control Around the World: Thermodynamic Recipe for Environmental Catastrophe. International Journal of Coal Geology 2004; 59(1):7–17.
5. Morris R. A Review of Experiences on the Use of Inert Gases in Mine Fires. Mining Science and Technology 1987; 6:37–69.
6. Greuer RE. Study of Mine Fire Fighting Using Inert Gases. USBF Contract Report No. 50231075. Bureau of Mines Open File Report 15-76. Bureau of Mines. Washington D.C.; 1974.
7. Harris L. The Use of Nitrogen to Control Spontaneous Combustion Heatings. The Mining Engineer 1981; 6:883–892.
8. Rehak TR, Bajpayee TS, Mowrey GL, Ingram DK. Flyrock Issues in Blasting. Proceedings of the 27th Annual Conference on Explosives and Blasting Technique 2001:165–175.
9. Verakis HC, Lobb TE. Blasting Accidents in Surface Mines, A Two Decade Summary. Proceedings of the 27th Annual Conference on Explosives and Blasting Technique 2001:145–152.
10. Singh PK, Roy SK, Sinha A. A New Blast Damage Index for the Safety of Underground Coal Mine Openings. Mining Technology 2003; 112(8):A97–A104.
11. Nundlall AR. Case Study: Lessons Learnt From Recent Flammable Gas Explosions in South African Hard Rock Mines. Journal of the Mine Ventilation Society of South Africa 2006; 59(2):64–69.
12. Tinko RJ. Methods to Determine the Status of Mine Atmosphere: An Overview. Journal of the Mine Ventilation Society of South Africa 2006; 59(2):46–55.
13. Koenning TH, Bruce WE. Mine Fire Indicators. Proceedings of the 3rd U.S. Mine Ventilation Symposium 1987:433–437.
14. Graham JI. The Origin of Blackdamp. Trans Inst Of Min Eng 1917–1918; LV:294–312.
15. Jones JE, Trickett JC. Some Observations on the Examination of Gases Resulting from Explosions in Colleries. Trans Inst Of Min Eng 1954–1955; 114:768–790.

Chapter 10
Safety in Offshore Industry

10.1 Introduction

Each year a vast sum of money is spent on offshore developments and offshore industry has become an important element of the industrial sector. Industry applies and develops leading-edge technology for drilling ever deeper and more efficiently to satisfy increasing demand for oil and gas. Some of the conditions under which Offshore installations sometimes must operate long distances from land, may encounter severe environments, and can include dangerous activities such as processing of explosive chemicals under high pressure [1]. Under such conditions safety has become a critical issue in the offshore industrial sector.

Over the years many accidents in offshore industry have occurred, resulting in many fatalities and millions of dollars in damages. Some of the deadliest accidents in the offshore industrial sector around the globe were the Piper Alpha platform accident in the United Kingdom in 1988, the Bohi No.2 jack-up accident in the Gulf of Bohi, China in 1979, the Enchova Central Platform accident in Brazil in 1988, the Mumbai High North Platform accident in India in 2005, and the Perforadora Central Usumacinta jack-up accident in Mexico in 2007. This chapter presents various important aspects of safety in offshore industry.

10.2 Offshore Risk Picture

The "risk picture" of the offshore industrial sector is a multi-faceted one. Establishment of an accurate "factual risk picture" is therefore generally a difficult task because of numerous factors. These include a basic uncertainty in extrapolating information from the past to present risk, inconsistency in recording and reporting incidents across the industry, changes in the regulatory regime and in the industry and its management systems, and different measures of risk [2].

Nonetheless, seven main contributors to individual risk have been identified and are as follows [2]:

- Process leaks that can ultimately develop into fires or explosions that may escalate
- Occupation-related accidents
- Ignited blow-outs with possible escalation
- Helicopter-related accidents on the platform itself
- Extreme environment-related loads
- Ramming by ships or other floating items
- Structural failures

Additional information on offshore risk picture is available in [2].

10.3 Offshore Worker Situation Awareness Concept

In many industrial sectors, there is a definite need to maintain the situation awareness of workers at a high level in order to insure the safety of their operations. Offshore industry is one the sectors for which this is particularly relevant. Situation awareness (SA) may simply be defined as the perception of the components in the environment within the framework of a volume of time and space, the understanding of their meanings, and the projection of their appropriate status in the near future [3].

Three main elements of SA are shown in Figure 10.1 [3].

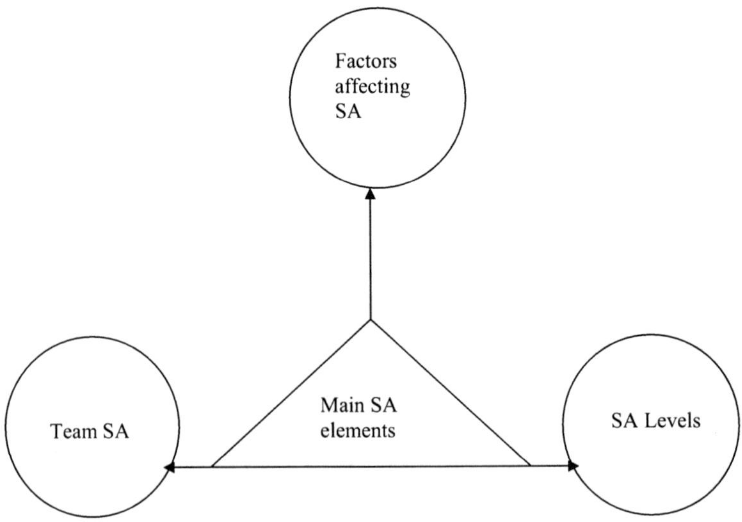

Figure 10.1 Primary situation awareness elements

SA levels are concerned with perception, comprehension, and projection. Perception calls for continual monitoring of the surrounding environment in order to encode sensory information as well as detecting changes in significant stimuli. Comprehension involves the combination, storage, interpretation, and retention of incoming information for the formation of a picture of the ongoing condition or situation whereby the significance of events/objectives is clearly understood. Finally, projection is the result of perception and comprehension and is concerned with the prediction of possible future events and states.

Team SA is basically concerned with teamwork, as the successful completion of a given task, e.g., a drilling task in the offshore industry, is totally dependent upon the crew collectively working together. Thus, it is essential for all team members to have a mutual understanding of the situation. In short, it may simply be stated as that all the team members should have a collective SA (this shared awareness is called team SA) [3, 4].

The factors affecting SA are stress and workload. Stress can lead to poor concentration or alertness because of overload on the individual's cognitive resources. Stressors can either be physical or psychological. Some examples of the physical stressors are noise, vibration, temperature, and pollution. Similarly, an example of the psychological stressors is anxiety. With respect to the SA model, stress can be seen to interfere with the primary perception of the situation under consideration [5].

Unusually low or high workloads are known to potentially impact human performance to a certain degree [6]. Low workload can lead to boredom with consequent inattentiveness, lower vigilance, and significantly reduced motivation. Less attention is being given to workplace situations or conditions which, in turn, can result in poor SA. On the other hand, high workloads can result in impairing worker's SA as they may not be fully aware of situational changes, and may make incorrect decisions on the basis of incomplete or wrong information. There is some evidence that increments in workload have detrimental impacts on the psychological well-being of offshore workers [3, 6, 7].

10.3.1 Studies and Their Results

A study of SA errors in offshore drilling incidents indicates that 66.7% of the incidents were perception-related, 20% comprehension-related, and 13.3% projection-related [3]. The breakdowns of the perception-related incidents were 26.8% (failure to monitor or observe data), 15.7% (hard to discriminate or detect data), 14.2% (misconception of data), 9.7% (data not available), and 0.1% (memory loss). Similarly, the breakdowns of the comprehension-related incidents were 11.1% (use of incorrect mental model), 6.7% (lack of/poor mental model), and 2.2% (over-reliance on default values). Finally, the breakdowns of the projection-related incidents were 13.2% (lack of/poor mental model) and 0.1% (over projection of current trends).

Table 10.1 A list of questions asked during interviews with offshore drilling-associated personnel

Question No.	Question
1	How is SA known in the offshore industry?
2	What factors affect the quality of a person's awareness?
3	What are the indicators of reduced awareness?
4	How can reduced awareness be improved?
5	How is team SA achieved?
6	What can be done to check the awareness of workers?

Another study was conducted which consisted of interviews with offshore drilling-associated personnel, and its aim was to understand how well the concept of SA is recognized within the offshore drilling industry. During the interviews six questions were asked. These are presented in Table 10.1 [3].

Three responses to question, "How is SA known in the offshore industry?" were safety awareness, safety accountability, and positional awareness. Twelve responses to question, "What factors affect the quality of a person's awareness?" were fatigue, home/family problems, stress and workload, experience (and new personnel), routine tasks/complacency, job prospects, having a near miss, weather/seasons, communication (good and bad), conflict, supervisory responsibility, and daydreaming.

Five responses to question, "What are the indicators of reduced awareness?" were reduction in communication, repetition of instructions, reduced work standards, character change, and an expressionless appearance. Ten responses to question, "How can reduced awareness be improved?" were communication, discussion of events, place them (personnel) in a different job, removal from the situation, alter the work level, alter the crew line-up, interaction, increase involvement in rig activities, training, and problem solving.

Eight responses to question, "How is team SA achieved?" were consistency, adaptability, co-operation, trust, time, understand capabilities and traits planning, experience, and increased interaction. Finally, two of the four responses to question, "What can be done to check the awareness of workers?" were risk assessments and constant assessments of surroundings. Additional information on these responses is available in [3].

10.4 Accident Reporting Approach in Offshore Industry

Generally, the accident reporting approach followed in offshore industry may be divided into two areas: minor accidents and serious incidents or accidents [8]. In the case of minor accidents, the investigators are the supervisors and safety officers, and usually have some degree of training. When the need arises, an investigation team is appointed for performing a more comprehensive investigation into the occurrence.

In the case of serious incidents or accidents, investigation teams fly to the offshore site from the onshore office along with government appointed accident inspectors. Usually, all accidents/incidents are documented and processed through appropriate channels, and the final copies of accident reports are distributed to all concerned bodies and authorities.

Normally, accident reporting forms contain information on items such as listed below [8]:

- Time, date, and location of occurrence
- Type of incident/accident: injury, property damage, poisonous or flammable substance leaks, material loss, disease, dangerous occurrences, process disruption, environmental harm, fire or explosion, near-misses, and hazards
- Personal details of individuals involved, including supervisor at the time of accident/incident
- Protective clothing being worn by all involved individuals
- Immediate and underlying causes
- Equipment failures
- The type of work being performed and experience of individuals involved
- Other people working in the surrounding area
- Permits being issued and procedures being followed
- Equipment being used, including safety devices and equipment
- Contributory factors (e.g., any hazards present, environmental conditions)

10.5 Offshore Accident Causes

A study of the accident reporting forms used by twenty five offshore companies in the United Kingdom revealed a large number of immediate causes for accidents [8]. Most of these immediate causes were operating without authority, use of defective equipment, failure to warn/secure, adjusted equipment in operation, workers under the influence of alcohol/drugs, incorrect speed, proper equipment not used, safety equipment/device made inoperable, improper lifting/loading, lack of attention/forgetfulness, equipment used improperly, work performed on unsafe or live equipment, and serviced equipment in operation.

In addition, there were a large number of underlying causes for accidents [8]. These underlying causes are grouped under two categories, termed personal factors and job factors. Four subcategories of the personal factors are shown in Figure 10.2. The elements in the stress subcategory were fatigue, monotony, health hazards, general stress, and frustration. There were nine elements of the capability subcategory: poor judgment, lack of physical capability, memory failure, lack of competence, inability to comprehend, lack of mental capability, judgment demands, perception demands, and concentration demands.

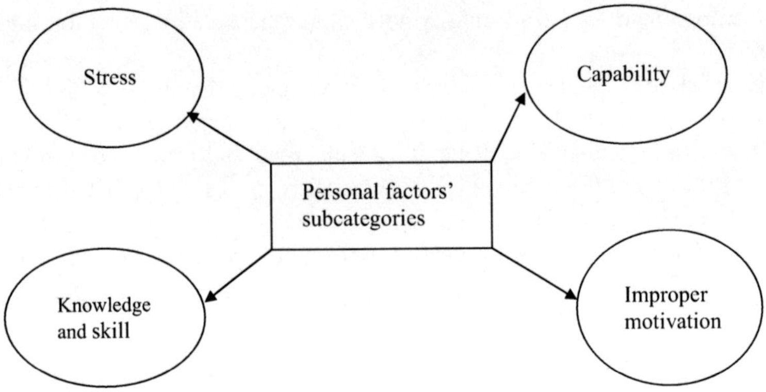

Figure 10.2 Subcategories of personal factors in offshore accident reporting forms

The improper motivation subcategory elements were peer pressure, lack of anticipation, inadequate thought and care, aggression, inappropriate attempt to save time, inattention, recklessness, horseplay, and attitude. Finally, there were a total of nine elements of the knowledge and skill subcategory: lack of experience, lack of awareness, lack of education, poor orientation, misunderstood directions, inadequate practice, poor training, lack of hands-on instructions, and lack of job instructions.

Four subcategories of the job factors are shown in Figure 10.3. The elements of the management subcategories were poor planning, communication, management practices, bad examples set by management, qualification and experience criteria, and management job knowledge. There were a total of eight elements of the organization subcategory: company policy, poor procedures, poor safety plan, poor staffing and resources, working hour policies, safety systems, adequacies of systems, and competence standards.

The supervision subcategory elements were inadequate work planning, unclear responsibilities, insufficient supervisory job knowledge, lack of inspections, in-

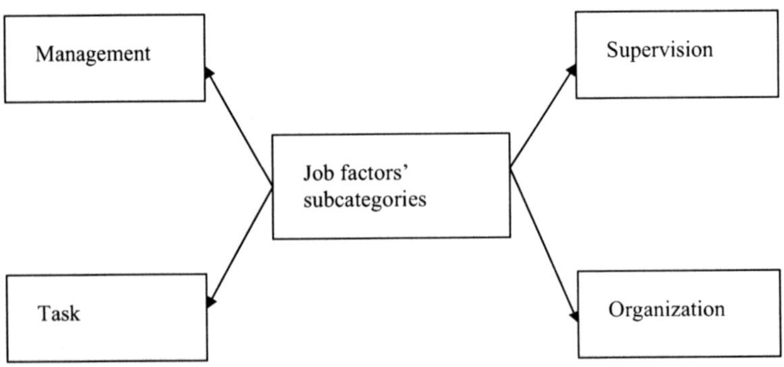

Figure 10.3 Subcategories of job factors in offshore accident reporting forms

complete instruction and training, inadequate discipline, improper production incentives, and poor supervisory examples. Finally, there were eight elements of the task subcategory: inadequate or no job description, conflicting goals, inadequate matching of individual to job task, confusing directions, time problems, failure in communication, inadequate work planning, and poor equipment selection. Additional information on all of the above subcategories is available in [8, 9].

10.6 Case Studies of Accidents in Offshore Industry

A large number of accidents in offshore industry have occurred over the years. Ten of the deadliest of these accidents occurred at or on the Piper Alpha (North Sea oil production platform), the Alexander L. Kielland (a Norwegian semi-submersible rig), the Usumcinta jack-up rig (Bay of Campeche, Gulf of Mexico), the Enchova Central Platform (Enchova Field, Brazil), the Seacrest Drillship (Platong Gas Field, Gulf of Thailand), the Ocean Ranger semi-submersible rig (Hibernia Field, North Atlantic), the Glomar Java Sea Drillship (South China Sea), the Bohai 2 jack-up (Gulf of Bohai, Off China), the C.P. Baker Drilling Barge (Eugene Island, Gulf of Mexico), and the Mumbai High North Platform (Mumbai (Bombay) High, Indian Ocean). Some of these accidents are described below, separately.

10.6.1 Piper Alpha Accident

Piper Alpha was a North Sea oil production platform located about 120 mi northeast of Aberdeen, Scotland. It was operated by Occidental Petroleum (Caledonia) Ltd. and it became operational in 1976. The platform was initially built to produce crude oil, but later also started to produce gas with the installation of gas conversion equipment. Piper Alpha produced oil and gas from 24 wells for delivery to the Flotta oil terminal on the Orkney Islands and to other installations through three separate pipelines. At the time of the disaster, the Piper Alpha platform produced approximately 10% of North Sea oil and gas [10–13].

On July 6, 1988 explosions and a resulting fire destroyed the platform and killed 167 persons [11, 12]. In terms of both lives lost and impact to offshore industry, this accident is considered to be the world's worst offshore industry disaster to date. A subsequent investigation into the disaster by the United Kingdom government highlighted many factors that contributed to the severity of the Piper Alpha incident. Two of these factors were the serious breakdown in the chain of command and lack of any proper communication to the platform's crew members, and the existence of fire walls and lack of blast walls [11–13]. More specifically for the latter factor, the existing fire walls predated the installation of the gas conversion equipment and were not properly upgraded to blast walls subsequent to the installation. The investigation resulted in a total of 106 recommendations for changes to existing North Sea safety-related procedures, and all of these recommendations were accepted by the offshore industry [11, 12]. Additional information on Piper Alpha accident is available in [10–13].

10.6.2 *Alexander L. Kielland Accident*

Alexander L. Kielland was a Norwegian semi-submersible rig in the Ekofisk oil field, Norwegian Continental Shelf, and was named after a Norwegian writer. It was located about 320 km east of Dundee, Scotland, U.K and was owned by the Stavanger Drilling Company of Norway. At the time of the disaster, the rig was hired by the United States company Phillips Petroleum.

After about 40 months of service, the Alexander Kielland was no longer used for drilling purposes, but it served as a so-called flotel (a floating hotel) for workers from the nearby Edda platform. On March 27, 1980 wind gusts of nearly 40 knots with waves up to 12 m high caused the rig to collapse into the North Sea, killing 123 off-duty workers.

A subsequent investigation by the Norwegian government into the disaster revealed that rig collapsed because of a fatigue crack in one of its six bracings (bracing D-6), which joined the collapsed D-leg to the rest of the rig in question [14]. In turn, the problem of the crack was traced to a rather small 6 mm fillet weld which connected a non load-bearing flange plate to this very D-6 bracing. Additional information on this accident is available in [14].

10.6.3 *Ocean Ranger Accident*

The Ocean Ranger was a semi-submersible mobile offshore drilling unit built for a Canadian company, Ocean Drilling and Exploration Company (ODECO), in 1976 by Mitsubishi Heavy Industry's Yard in Hiroshima, Japan. The drilling rig was composed of a square upper hull supported by 8 vertical columns.

On February 15, 1982 the Ocean Ranger was drilling an exploration well about 166 mi east of St. John's, Newfoundland, Canada and sank with 84 crew members on board. There were no survivors.

A subsequent investigation by a Canadian Royal Commission into the disaster concluded that the rig had some design and construction flaws, particularly in the blast control room [15]. In addition, the commission highlighted that the crew members lacked proper safety training, equipment, and survival suits. Additional information on this accident is available in [15, 16].

10.6.4 Glomar Java Sea Drillship Accident

The Glomar Java Sea Drillship was designed by Global Marine, Inc., and constructed by the Livingston Shipbuilding Company of Orange, Texas in 1975. It was built for drilling wells down to about 25,000 ft in water depths of approximately up to 1,000 ft.

On October 25, 1983, the United States drillship capsized and sank in the South China Sea about 80 nautical miles east of the Socialist Republic of Vietnam and 63 nautical miles Southwest of Hainan Island, China. All 81 persons onboard the drillship were killed.

A subsequent investigation by the National Transportation Safety Board concluded that the most likely cause for the capsizing and sinking of the Glomar Java Sea Drillship during typhoon Lex was the flooding of its starboard tanks 6 and 7 through a hull fracture [17]. Additional information on this accident is available in [17].

10.6.5 Enchova Central Platform Accident

The Enchova Central Platform was located in the Campos Basin near Rio de Janeiro, Brazil and it was run by the Brazilian company Petrobras. On August 16, 1984, a blow-out resulted in an explosion and fire. Although most of the personnel were evacuated safely by lifeboat or helicopter, 42 fatalities occurred during the platform evacuation process.

In this case, the most serious incident occurred when a lifeboat's lowering mechanism failed. Consequently, the life-boat was left suspended in a vertical position until the stern support broke and the life boat fell about 20m to the sea and killed 36 of its occupants. Another 6 persons died when they jumped about 40 m from the platform to the sea below. Additional information on this accident is available in [18].

10.6.6 Mumbai High North Platform Accident

The Mumbai High field is India's largest oil and gas field and it was discovered in 1974. The field is located about 100 mi west of the Mumbai coast in the Arabian Sea. The Mumbai High North platform was built in 1981 and was an oil and natural gas processing complex with a capacity of 80,000 barrels per day of crude oil production.

The platform was a 7-story high steel structure containing five gas export risers and ten fluid import risers located just outside its jacket. On July 27, 2005, a multipurpose support vessel collided with the platform that severed at least one gas riser and caused a major fire. The platform was destroyed within hours and 22 persons were killed. Additional information on this accident is available in [19, 20].

10.6.7 Bohai 2 Jack-Up Accident

The Bohai No. 2 jack-up was located in the Gulf of Bohai off China, and it was operated by the Ocean Oil Company, China Petroleum Department. It was a Self-Elevating Drilling Unit that sank under tow after encountering a storm with force 10 winds on November 25, 1979. This incident resulted in the deaths of 72 of the 74 persons on board.

A subsequent investigation into the accident attributed various causes for its severity. Some of these causes are listed below [21]:

- Failure to correctly stow deck equipment prior to towing
- Failure to follow standard tow procedures with respect to weather
- Poor training of crew members with respect to the use of life saving equipment
- Poor emergency evacuation procedures

Additional information on this accident is available in [21–23].

10.7 Problems

1. Write an essay on safety in offshore industry.
2. Discuss the offshore risk picture by stating the main contributors to individual risk.
3. Discuss the situation awareness (SA) concept.
4. Discuss an accident reporting approach usually followed in the offshore industrial sector.
5. List at least ten items on which accident reporting forms used in offshore industry contain information.
6. What are the typical causes of offshore accidents?
7. What are the two main categories of underlying causes for accidents in offshore industry?
8. Discuss the Piper Alpha accident.
9. Compare the Piper Alpha accident with the Alexander L. Kielland accident.
10. Discuss the following three accidents:
    - Glomar Java Sea Drillship accident
    - Ocean Ranger accident
    - Mumbai High North Platform accident

References

1. Falker III JM, Nickerson W. A New Direction for Safety Policy: The Offshore Oil Industry and Safety Regulation of Technology. Technology in Society 1996; 18(4):503–510.
2. Tveit OJ. Safety Issues on Offshore Process Installation: An Overview. J Loss Prev Process Ind 1994; 7(4):267–272.
3. Sneddon A, Mearns K, Flin R. Situation Awareness and Safety in Offshore Drill Crews. Cognition, Technology, and Work 2006; 8(4):255–267.
4. Shu Y, Furuta K. An Inference Method of Team Situation Awareness Based on Mutual Awareness. Cognition, Technology, and Work 2005; 7(4):272–287.
5. Endsley M. Toward a Theory of Situation Awareness in Dynamic Systems. Human Factors 1995; 37(1):32–64.
6. Parkes K. Psychosocial Aspects of Stress, Health and Safety on North Sea Installations. Scandinavian Journal of Work, Environment, and Health 1998; 24(5): 321–333.
7. Sutherland K, Flin R. Stress at Sea: A Review of Working Conditions in the Offshore Oil and Fishing Industries. Work Stress 1989; 3:269–285.
8. Gordon RPE.The Contribution of Human Factors to Accidents in the Offshore Oil Industry. Reliability Engineering and System Safety 1998; 61:95–108.
9. Bird FE, Germain GL. Practical Loss Control Leadership: The Conservation of People, Property, Process, and Profits. Logansville, Georgia: Institute Publishing (ILCI); 1989.
10. Petrie JR. Piper Alpha Technical Investigation Interim Report. Petroleum Engineering Division. Department of Energy. London, U.K.; 1988.
11. Pate-Cornell ME. Risk Analysis and Risk Management for Offshore Platforms: Lessons from the Piper Alpha Accident. Journal of Offshore Mechanics and Arctic Engineering 1993; 115(1):179–190.
12. Pate-Cornell ME. Learning from the Piper Alpha Accident: Analysis of Technical and Organizational Factors. Risk Analysis 1993; 13(2):215–232.
13. Hull AM, Alexander DA, Klein S. Survivors of the Piper Alpha Oil Platform Disaster: Long-Term Follow up Study. The British Journal of Psychiatry 2002;181:433–438.
14. Alexander L. Kielland Accident. Report of a Norwegian Public Commission Appointed by Royal Decree of March 28, 1980, Norwegian Ministry of Justice Police, Oslo, Norway, March 1981 (ISBN B0000ED27N).
15. Report of the Royal Commission on the Ocean Ranger Marine Disaster. Report ISBN No. 0660116820, Government of Canada, Ottawa, Canada; 1984.
16. Mobile Offshore Drilling Unit (MODU) Ocean Ranger (O.N.615641) Capsizing and Sinking in the Atlantic Ocean. Report No. USCG 0001 HQS 82, United States Coast Guard, Department of Transportation, Washington, D.C.; 1983.
17. Capsizing and Sinking of the United States Drillship Glomar Java Sea. Report No. NTSB-MAR-84-08. National Transportation Safety Board. Washington, D.C.; 1984.
18. One Hundred Largest Losses. Marsh Risk Consulting. Available from: URL: http://www.marshriskconslting.com/st/PSEV-C-352-NR-304.htm.
19. Mumbai High North. Available from: URL: http://home.versatel.nl/the-sims/rig/mhn.htm.
20. Riser Safety in UK Waters – Lessons from Mumbai High North Disaster. Report No. SPC/Technical/OSD/33. Hazardous Installations Directorate. Health and Safety Executive. London: May 2006.
21. Bohai 2 Jack-Up. Available from: URL: http://home.versatel.nl/the-sims/rig/bohai2.htm.
22. Santos RS, Feijo LP. Safety Challenges Associated with Deepwater Concepts Utilized in the Offshore Industry. Proceedings of the 9th International Symposium on Maritime Health 2007; 6:1–8.
23. Santos RS, Feijo LP. Deepwater Safety Challenges to Consider in a Fast-Paced Development Environment. Offshore 2008; 68(3):1–5.

Chapter 11
Mathematical Models for Performing Safety Analysis in Mines

11.1 Introduction

Mathematical modeling is a technique that is widely used to perform various types of analysis in the area of engineering. In such models, system components are represented by idealized elements assumed to have all the representative characteristics of real-life components, and whose behavior is possible to be described by equations. However, the degree of realism of a mathematical model depends on the assumptions imposed upon it.

Over the years, a large number of mathematical models have been developed to study the reliability, maintainability, and safety of engineering systems. Many of these models were developed using stochastic processes including the Markov approach [1–3]. Although the effectiveness of such models can vary quite considerably from one application area to another, some of these models are being used quite successfully for studying various types of real-life problems in industry [4, 5]. Thus, some of these mathematical models can also be used to study safety-related problems in the area of mining. This chapter presents the mathematical models considered useful to perform various types of safety-related analysis in mines.

11.2 Model I

This mathematical model represents a mining system having three distinct states: working normally, failed unsafely, and failed safely. The failed system (i.e., unsafely or safely) is repaired to its normal working state. The mining system state space diagram is shown in Figure 11.1 [7].

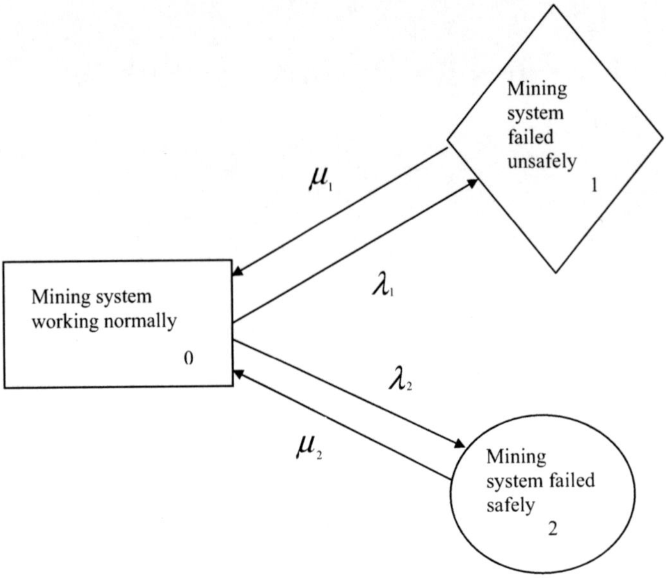

Figure 11.1 Mining system state space diagram

The numerals in the box, circle, and diamond denote system states. The following assumptions are associated with this model:

- All failures occur independently.
- The mining system can fail either unsafely or safely.
- The mining system unsafe and safe failure rates are constant.
- The failed mining system repair rates are constant.
- The repaired mining system is as good as new.

The following symbols are associated with this model:

j is the jth state of the mining system, where $j = 0$ means the system is working normally, $j = 1$ means the system failed unsafely, and $j = 2$ means the system failed safely.

$P_j(t)$ is the probability that the system is in state j at time t, for $j = 0$, 1,2.

t is time.

λ_j is the mining system jth failure rate, where $j = 1$ means unsafe and $i = 2$ means safe.

μ_j is the failed mining system's jth repair rate, where $j = 1$ means from an unsafe failed state, and $j = 2$ means from a safe failed state.

Using the Markov method described in Chapter 4 and [1, 7], we write down the following set of differential equations for Figure 11.1:

$$\frac{dP_0(t)}{dt} + (\lambda_1 + \lambda_2)P_0(t) = \mu_1 P_1(t) + \mu_2 P_2(t) \tag{11.1}$$

$$\frac{dP_1(t)}{dt} + \mu_1 P_1(t) = \lambda_1 P_0(t) \tag{11.2}$$

$$\frac{dP_2(t)}{dt} + \mu_2 P_2(t) = \lambda_2 P_0(t) \tag{11.3}$$

At time $t=0$, $P_0(0)=1$, and $P_1(0)=P_2(0)=0$.

By solving Equations 11.1–11.3, we get

$$P_0(t) = \frac{\mu_1\mu_2}{Y_1Y_2} + \left[\frac{(Y_1+\mu_2)(Y_1+\mu_1)}{Y_1(Y_1-Y_2)}\right]e^{Y_1t} - \left[\frac{(Y_2+\mu_2)(Y_2+\mu_1)}{Y_2(Y_1-Y_2)}\right]e^{Y_2t} \tag{11.4}$$

where

$$Y_1, Y_2 = \frac{-B \pm \sqrt{B^2 - 4(\mu_2\mu_1 + \lambda_2\mu_1 + \lambda_1\mu_1)}}{2}$$

$$B = \mu_2 + \mu_1 + \lambda_1 + \lambda_2$$

$$Y_1Y_2 = \mu_1\mu_2 + \lambda_2\mu_1 + \lambda_1\mu_2$$

$$Y_1 + Y_2 = -(\mu_2 + \mu_1 + \lambda_2 + \lambda_1)$$

$$P_1(t) = \frac{\lambda_1\mu_2}{Y_1Y_2} + \left[\frac{(\lambda_1Y_1 + \lambda_1\mu_2)}{Y_1(Y_1-Y_2)}\right]e^{Y_1t} - \left[\frac{(\mu_2+Y_2)\lambda_1}{Y_2(Y_1-Y_2)}\right]e^{Y_2t} \tag{11.5}$$

$$P_2(t) = \frac{\lambda_2\mu_1}{Y_1Y_2} + \left[\frac{(\lambda_2Y_1 + \lambda_2\mu_1)}{Y_1(Y_1-Y_2)}\right]e^{Y_1t} - \left[\frac{(\mu_1+Y_2)\lambda_2}{Y_2(Y_1-Y_2)}\right]e^{Y_2t} \tag{11.6}$$

Equations 11.5 and 11.6 give the probability of the mining system failing unsafely and safely, respectively, when subjected to the repair process.

As time t becomes very large, the mining system steady-state probability of failing unsafely using Equation 11.5 is

$$P_1 = \lim_{t\to\infty} P_1(t) = \frac{\lambda_1\mu_2}{Y_1Y_2} \tag{11.7}$$

Similarly, as time t becomes very large, the mining system steady state probability of failing safely using Equation 11.6 is

$$P_2 = \lim_{t\to\infty} P_2(t) = \frac{\lambda_2\mu_1}{Y_1Y_2} \tag{11.8}$$

By setting $\mu_1=\mu_2=0$ in Equations 11.1–11.3 and then solving the resulting equations, we obtain

$$P_0(t) = e^{-(\lambda_1+\lambda_2)t} \tag{11.9}$$

$$P_1(t) = \frac{\lambda_1}{\lambda_1 + \lambda_2}\left[1 - e^{-(\lambda_1+\lambda_2)t}\right] \tag{11.10}$$

$$P_2(t) = \frac{\lambda_2}{\lambda_1 + \lambda_2}\left[1 - e^{-(\lambda_1+\lambda_2)t}\right] \tag{11.11}$$

Equation 11.9 gives the mining system reliability at time t. In contrast, Equations 11.10 and 11.11 give the probability of the mining system failing unsafely and safely at time t, respectively.

By integrating Equation 11.9 over the time interval [0, ∞], we obtain the following expression for the mining system mean time to failure [8]:

$$\begin{aligned} MTTF_{ms} &= \int_0^\infty P_0(t)\,dt \\ &= \int_0^\infty e^{-(\lambda_1+\lambda_2)t}\,dt \\ &= \frac{1}{\lambda_1 + \lambda_2} \end{aligned} \tag{11.12}$$

where $MTTF_{ms}$ is the mining system mean time to failure.

Example 11.1

A mining equipment can fail safely or unsafely and its constant failure rates are 0.0005 failures per hour and 0.0001 failures per hour, respectively. Calculate the probability of the mining equipment failing unsafely during a 100-h mission.

By substituting the specified data values into Equation 11.10, we get

$$\begin{aligned} P_1(100) &= \frac{0.0001}{0.0001 + 0.0005}\left[1 - e^{-(0.0001+0.0005)(100)}\right] \\ &= 0.0097 \end{aligned}$$

Thus, the probability of the mining equipment failing unsafely during the specified mission period is 0.0097.

11.3 Model II

This mathematical model represents a mining system having three distinct states: working normally, working unsafely, and failed. The system is repaired from unsafe working state and fully failed state and its transition diagram is shown in Figure 11.2. The numerals in rectangles and the circle denote mining system states.

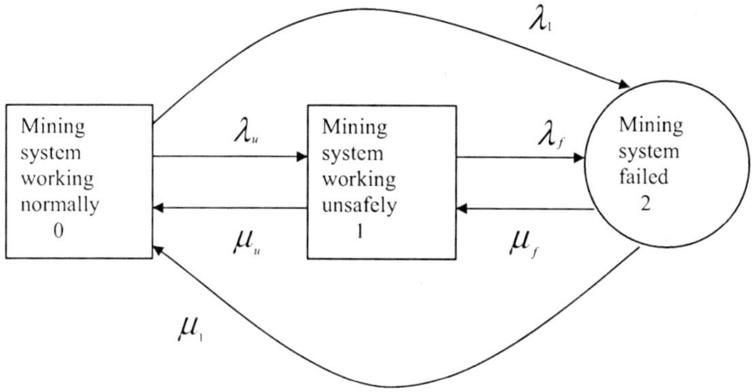

Figure 11.2 Mining system transition diagram

The following assumptions are associated with this model:

- Failures occur independently.
- The mining system failure and repair rates are constant.
- The repaired mining system is as good as new.

The following symbols are associated with this model:

j is the jth state of the mining system, where $j = 0$ means the mining system is working normally, $j = 1$ means the mining system is working unsafely, and $j = 2$ means the mining system failed.

t is time.

$P_j(t)$ is the probability that the mining system is in state j at time t, for $j = 0, 1, 2$.

λ_j is the mining system jth constant failure rate, where $j = 1$ means from state 0 to state 2, $j = u$ means from state 0 to state 1, $j = f$ means from state 1 to state 2.

μ_j is the mining system jth constant repair rate, where $j = 1$ means from state 2 to state 0, $j = u$ means from state 1 to state 0, $j = f$ means from state 2 to state 1.

Using the Markov approach described in Chapter 4 and [1, 7], we write down the following system of differential equations for Figure 11.2:

$$\frac{\mathrm{d}P_0(t)}{\mathrm{d}t} + (\lambda_u + \lambda_1)P_0(t) = \mu_u P_1(t) + \mu_1 P_2(t) \tag{11.13}$$

$$\frac{\mathrm{d}P_1(t)}{\mathrm{d}t} + (\mu_u + \lambda_f)P_1(t) = \mu_f P_2(t) + \lambda_u P_0(t) \tag{11.14}$$

$$\frac{\mathrm{d}P_2(t)}{\mathrm{d}t} + (\mu_1 + \mu_f)P_2(t) = \lambda_f P_1(t) + \lambda_1 P_0(t) \tag{11.15}$$

At time $t = 0$, $P_0(0) = 1$, and $P_1(0) = P_2(0) = 0$.

For a very large t, by solving Equations 11.13–11.15, we obtain the following steady state probabilities [8]:

$$P_0 = \frac{(\mu_1 + \mu_f)(\mu_u + \lambda_f) - \lambda_f \mu_f}{B} \tag{11.16}$$

where

$$B = (\mu_1 + \mu_f)(\mu_u + \lambda_u + \lambda_f) + \lambda_1(\mu_u + \lambda_f) + \lambda_1 \mu_f + \lambda_u \lambda_f - \lambda_f \mu_f$$

$$P_1 = \frac{\lambda_u(\mu_1 + \mu_f) + \lambda_1 \mu_f}{B} \tag{11.17}$$

$$P_2 = \frac{\lambda_1 \lambda_f + \lambda_1(\mu_u + \lambda_f)}{B} \tag{11.18}$$

where

P_0, P_1, and P_2 are the steady state probabilities of the mining system being in states 0, 1, and 2, respectively.

By setting $\mu_f = \mu_1 = 0$ in Equations 11.13–11.15 and solving the resulting equations, we obtain the following expression for the mining system reliability:

$$\begin{aligned} R_{ms}(t) &= P_0(t) + P_2(t) \\ &= (X_1 + Y_1)e^{r_1 t} + (X_2 + Y_2)e^{r_2 t} \end{aligned} \tag{11.19}$$

where

$R_{ms}(t)$ is the mining system reliability at time t.

$$r_1 = \frac{-L_1 + \sqrt{L_1^2 - 4L_2}}{2}$$

$$r_2 = \frac{-L_1 - \sqrt{L_1^2 - 4L_2}}{2}$$

$$L_1 = \mu_u + \lambda + \lambda_u + \lambda_f$$

$$L_2 = \lambda_1 \mu_u + \lambda_1 \lambda_f + \lambda_u \lambda_f$$

$$X_1 = \frac{r_1 + \mu_u + \lambda_f}{(r_1 - r_2)}$$

$$X_2 = \frac{r_2 + \mu_u + \lambda_f}{(r_2 - r_1)}$$

$$Y_1 = \frac{\lambda_u}{(r_1 - r_2)}$$

$$Y_2 = \frac{\lambda_u}{(r_2 - r_1)}$$

By integrating Equation 11.19 over the time interval [0, ∞], we obtain the following equation for the mining system mean time to failure:

$$MTTF_{ms} = \int_0^\infty R_{ms}(t)\,dt = \left[\frac{(X_1 + Y_1)}{r_1} + \frac{(X_2 + Y_2)}{r_2}\right] \tag{11.20}$$

Example 11.2

Assume that the following values of parameters associated with a mining system are specified:

$\lambda_1 = 0.005$ failures per hour
$\lambda_u = 0.002$ failures per hour
$\lambda_f = 0.001$ failures per hour
$\mu_1 = 0.006$ repairs per hour
$\mu_u = 0.003$ repairs per hour
$\mu_f = 0.004$ repairs per hour

Calculate the steady state probability of the mining system working unsafely by using Equation 11.17.

By substituting the specified data values into Equation 11.17, we get

$$P_1 = \frac{(0.002)(0.006 + 0.004) + (0.005)(0.004)}{B} = 0.392$$

where

$$B = (0.006 + 0.004)(0.003 + 0.002 + 0.001) + (0.005)(0.003 + 0.001) + (0.005)(0.004) + (0.002)(0.001) - (0.001)(0.004)$$

Thus, the steady state probability of the mining system working unsafely is 0.3921.

11.4 Model III

This model is concerned with predicting the reliability of a worker performing a mining task. In this case, the reliability is expressed by [3, 8, 9]:

$$R_w(t) = \exp - \left[-\int_0^t \lambda_w(x)\,dx \right] \tag{11.21}$$

where

$R_w(t)$ is the worker reliability at time t.
$\lambda_w(x)$ is the worker instantaneous error rate.

It is to be noted that in Equation 11.21, the time to human error can follow any time continuous probability distribution (e.g., Weibull, gamma, and exponential).

Example 11.3

A worker is performing a time continuous mining task and his/her error rate is 0.005 errors per hour (i.e., the worker's times to error are exponentially distributed). Calculate the worker's reliability for an 8-h mission.

By substituting the specified data values into Equation 11.21, we obtain

$$\begin{aligned} R_w(8) &= \exp\left[-\int_0^8 (0.005)\,dx\right] \\ &= \exp[-(0.005)(8)] \\ &= 0.9608 \end{aligned}$$

Thus, the worker's reliability during the specified mission time is 0.9608.

11.5 Model IV

This mathematical model represents a mining situation where a worker performs a time-continuous task under fluctuating environments (i.e., normal and abnormal) [10]. The worker can commit an error in either normal or abnormal environments.

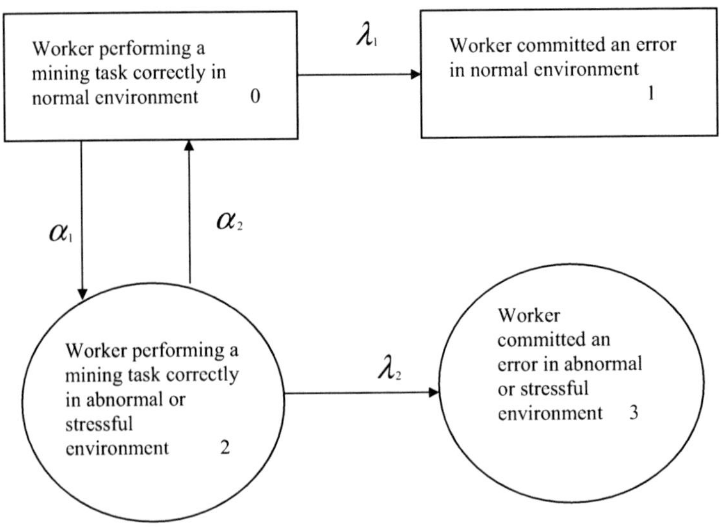

Figure 11.3 State space diagram for the worker performing in fluctuating environments

The state space diagram for the worker performing a time-continuous mining task under fluctuating environments is shown in Figure 11.3.

The following assumptions are associated with this model:

- The worker is performing a time-continuous task.
- All errors occur independently.
- Worker error rates are constant.
- The rate of changing environment from normal to abnormal (or stressful) or *vice versa* is constant.
- Numerals in Figure 11.3 boxes and circles denote corresponding states (e.g., worker committed an error in normal environment: 1).

The following symbols are associated with this model (i.e., Figure 11.3):

$P_j(t)$ is the probability of the worker being in state j at time t, for $j = 0, 1, 2, 3$.
α_1 is the constant transition rate from normal environments to abnormal or stressful environments.
α_2 is the constant transition rate from abnormal or stressful environments to normal environments.
λ_1 is the constant error rate of the worker performing a mining task in normal environments.
λ_2 is the constant error rate of the worker performing a mining task in abnormal or stressful environments.

Using the Markov method described in Chapter 4 and [1, 3, 10], we write down the following set of differential equations for Figure 11.3:

$$\frac{dP_0(t)}{dt} + (\lambda_1 + \alpha_1)P_0(t) = \alpha_2 P_2(t) \tag{11.22}$$

$$\frac{dP_1(t)}{dt} = \lambda_1 P_0(t) \tag{11.23}$$

$$\frac{dP_2(t)}{dt} + (\lambda_2 + \alpha_2)P_2(t) = \alpha_1 P_0(t) \tag{11.24}$$

$$\frac{dP_3(t)}{dt} = \lambda_2 P_2(t) \tag{11.25}$$

At time $t = 0$, $P_0(0) = 1$ and $P_1(0) = P_2(0) = P_3(0) = 0$.

By solving Equations 11.22–11.25, we obtain

$$P_0(t) = (x_1 - x_2)^{-1}\left[(x_2 + \lambda_2 + \alpha_2)e^{x_2 t} - (x_1 + \lambda_2 + \alpha_2)e^{x_1 t}\right] \tag{11.26}$$

where

$$x_1 = \frac{-b_1 + \sqrt{b_1^2 - 4b_2}}{2}$$

$$x_2 = \frac{-b_1 - \sqrt{b_1^2 - 4b_2}}{2}$$

where

$$b_1 = \lambda_1 + \lambda_2 + \alpha_1 + \alpha_2$$

$$b_2 = \lambda_1(\lambda_2 + \alpha_2) + \alpha_1\lambda_2$$

$$P_1(t) = b_4 + b_5 e^{x_2 t} - b_6 e^{x_1 t} \tag{11.27}$$

where

$$b_3 = \frac{1}{x_2 - x_1}$$

$$b_4 = \frac{\lambda_1(\lambda_2 + \alpha_2)}{x_1 x_2}$$

$$b_5 = b_3(\lambda_1 + b_4 x_1)$$

$$b_6 = b_3(\lambda_1 + b_4 x_2)$$

$$P_2(t) = \alpha_1 b_3 (\mathrm{e}^{x_2 t} - \mathrm{e}^{x_1 t}) \tag{11.28}$$

$$P_3(t) = b_7\left[(1 + b_3)(x_1 \mathrm{e}^{x_2 t} - x_2 \mathrm{e}^{x_1 t})\right] \tag{11.29}$$

where

$$b_7 = \frac{\lambda_2 \alpha_1}{x_1 x_2}$$

The worker reliability in fluctuating environment is expressed by

$$R_w(t) = P_0(t) + P_2(t) \tag{11.30}$$

where

$R_w(t)$ is the worker reliability of performing a mining task in fluctuating environments at time t.

The mean time to worker error is given by [10]

$$MTTWE = \int_0^{\infty} R_w(t)dt = \frac{(\lambda_2 + \alpha_1 + \alpha_2)}{b_2} \tag{11.31}$$

where $MTTWE$ is the mean time to worker error.

Example 11.4

A worker is performing a mining task in fluctuating (i.e., normal and stressful) environments and his/her error rates are 0.001 errors per hour and 0.008 errors per hour, respectively. The transition rates from normal to stressful environments and vice versa are 0.004/h and 0.003/h, respectively. Calculate the mean time to worker error.

By substituting the given data values into Equation 11.31, we obtain

$$MTTWE = \frac{(0.008+0.004+0.003)}{(0.001)(0.008+0.003)+(0.004)(0.008)}$$
$$= 348.8\,\text{h}$$

Thus, mean time to worker error is 348.8 h.

11.6 Model V

This mathematical model represents a worker performing a time continuous mining task subjected to critical (unsafe) and noncritical (safe) errors. In other words, the errors committed by the worker are grouped under two categories: critical (unsafe) and noncritical (safe). The state space diagram for the worker performing a time-continuous mining task is shown in Figure 11.4 [6].

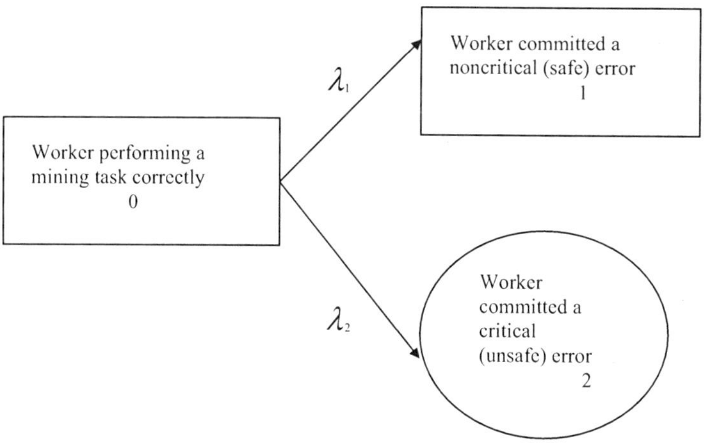

Figure 11.4 State space diagram for the worker performing a mining task

The numerals in boxes and the circle denote the states of the worker. The following assumptions are associated with this model:

- Worker critical and noncritical error rates are constant.
- All errors occur independently.
- The worker performs a time-continuous task.

The following symbols are associated with this model (i.e., Figure 11.4):

i is the ith state of worker performing a mining task, where $i = 0$ means that the worker is performing a mining task correctly, $i = 1$ means that the worker committed a noncritical (safe) error, $i = 2$ means that the worker committed a critical (unsafe) error.

$P_i(t)$ is the probability of the worker being in state i at time t; for $i = 0, 1, 2$.

λ_1 is the worker noncritical (safe) error rate.

λ_2 is the worker critical (unsafe) error rate.

Using the Markov method described in Chapter 4 and [1, 3], we write down the following system of differential equations for Figure 11.4:

$$\frac{\mathrm{d}P_0(t)}{\mathrm{d}t} + (\lambda_1 + \lambda_2)P_0(t) = 0 \tag{11.32}$$

$$\frac{\mathrm{d}P_1(t)}{\mathrm{d}t} - \lambda_1 P_0(t) = 0 \tag{11.33}$$

$$\frac{\mathrm{d}P_2(t)}{\mathrm{d}t} - \lambda_2 P_0(t) = 0 \tag{11.34}$$

At time $t = 0$, $P_0(0) = 1$, $P_1(0) = 0$, and $P_2(0) = 0$.

By solving Equations 11.32–11.34, we obtain

$$P_0(t) = \mathrm{e}^{-(\lambda_1 + \lambda_2)t} \tag{11.35}$$

$$P_1(t) = \frac{\lambda_1}{\lambda_1 + \lambda_2}\left[1 - \mathrm{e}^{-(\lambda_1 + \lambda_2)t}\right] \tag{11.36}$$

$$P_2(t) = \frac{\lambda_2}{\lambda_1 + \lambda_2}\left[1 - \mathrm{e}^{-(\lambda_1 + \lambda_1)t}\right] \tag{11.37}$$

The worker reliability is given by

$$\begin{aligned} R_w(t) &= P_0(t) \\ &= \mathrm{e}^{-(\lambda_1 + \lambda_2)t} \end{aligned} \tag{11.38}$$

where

$R_w(t)$ is the worker reliability at time t.

The mean time to worker error (*MTTWE*) is expressed by [3, 6]:

$$MTTWE = \int_0^\infty R_w(t)\,dt = \int_0^\infty e^{-(\lambda_1+\lambda_2)t}\,dt = \frac{1}{\lambda_1+\lambda_2} \tag{11.39}$$

Example 11.5

Assume that a worker is performing a mining task and his/her critical (unsafe) and noncritical (safe) error rates are 0.004 errors per hour and 0.008 errors per hour, respectively. Calculate the probability of the worker committing a critical (unsafe) error during a 7-h work period.

By substituting the given data values into Equation 11.37, we get

$$P_2(7) = \frac{0.004}{(0.008+0.004)}\left[1-e^{-(0.008+0.004)(7)}\right] = 0.0268$$

Thus, the probability of the worker committing a critical (unsafe) error during the specified period is 0.0268.

11.7 Model VI

This mathematical model represents mining equipment that can either fail safely or unsafely. The failed equipment is taken to the repair workshop [3, 11]. The model state space diagram is shown in Figure 11.5.

The numerals in circles denote the equipment states. The following assumptions are associated with this model:

- Failures occur independently.
- Equipment failure rates and the rates of taking the failed equipment to the repair workshop are constant.

The following symbols are associated with this model:

j is the jth state of the mining equipment, where $j = 0$ means that the mining equipment operating normally, $j = 1$ means that the mining equipment failed unsafely, $j = 2$ means that the mining equipment failed safely, and $j = 3$ means that the mining equipment is in the repair workshop.

$P_j(t)$ is the probability that the mining equipment is in state j at time t, for $j=0, 1, 2, 3$.

λ_s is the mining equipment constant safe failure rate.

λ_u is the mining equipment constant unsafe failure rate.

λ_1 is the constant rate of taking the unsafely failed mining equipment to the repair workshop.

λ_2 is the constant rate of taking the safely failed mining equipment to the repair workshop.

Using the Markov method described in Chapter 4 and [1, 3], we write down the following set of differential equations for Figure 11.5:

$$\frac{dP_0(t)}{dt} + (\lambda_s + \lambda_u)P_0(t) = 0 \tag{11.40}$$

$$\frac{dP_1(t)}{dt} + \lambda_1 P_1(t) = \lambda_u P_0(t) \tag{11.41}$$

$$\frac{dP_2(t)}{dt} + \lambda_2 P_2(t) = \lambda_s P_0(t) \tag{11.42}$$

$$\frac{dP_3(t)}{dt} = \lambda_1 P_1(t) + \lambda_2 P_2(t) \tag{11.43}$$

At time $t=0$, $P_0(0)=1$, and $P_1(0)=P_2(0)=P_3(0)=0$.

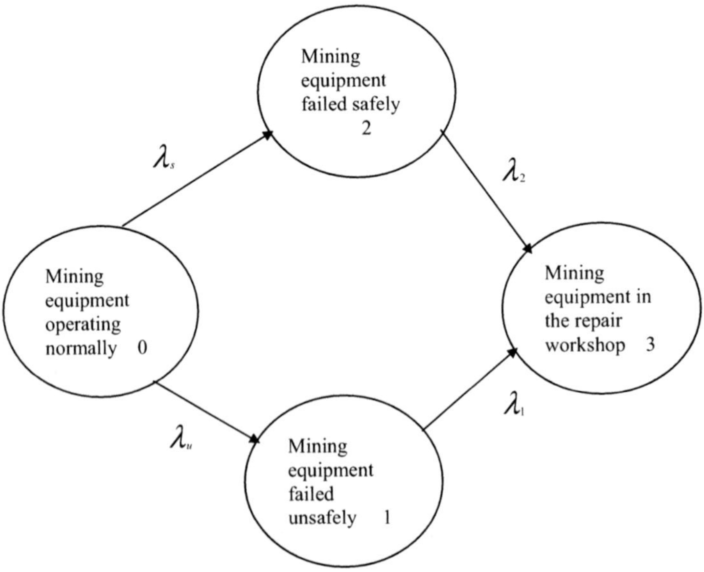

Figure 11.5 State space diagram for a mining equipment failing safely or unsafely and taken to repair workshop

By solving Equations 11.40–11.43, we get

$$P_0(t) = \mathrm{e}^{-At} \tag{11.44}$$

where

$$A = \lambda_s + \lambda_u$$

$$P_1(t) = B(\mathrm{e}^{-AT} - \mathrm{e}^{-\lambda_1 t}) \tag{11.45}$$

where

$$B = \lambda_u / (\lambda_1 - A)$$

$$P_2(t) = C(\mathrm{e}^{-At} - \mathrm{e}^{-\lambda_2 t}) \tag{11.46}$$

where

$$C = \lambda_s / (\lambda_2 - A)$$

$$P_3(t) = 1 + B\mathrm{e}^{-\lambda_1 t} + C\mathrm{e}^{-\lambda_2 t} + D\mathrm{e}^{-At} \tag{11.47}$$

where

$$D = -(B\lambda_1 + C\lambda_2) / A$$

The mining equipment reliability is given by

$$R_{me}(t) = P_0(t) = \mathrm{e}^{-At} \tag{11.48}$$

where

$R_{me}(t)$ is the mining equipment reliability at time t.

The mining equipment mean time to failure ($MTTF_{me}$) is given by [3]:

$$\begin{aligned} MTTF_{me} &= \int_0^\infty R_{me}(t)\,\mathrm{d}t \\ &= \int_0^\infty \mathrm{e}^{-At}\,\mathrm{d}t \\ &= \frac{1}{A} \\ &= \frac{1}{\lambda_s + \lambda_u} \end{aligned} \tag{11.49}$$

Example 11.6

Assume that the constant safe and unsafe failure rates of a piece of mining equipment are 0.006 failures per hour and 0.001 failures per hour, respectively. Calculate the mining equipment reliability during a 12-h mission.

By substituting the given data values into Equation 11.48, we get

$$R_{me}(12) = e^{-(0.006+0.001)(12)}$$
$$= 0.9194$$

Thus, the mining equipment reliability during the specified period is 0.9194.

11.8 Model VII

This mathematical model is basically same as the previous mathematical model (Model VI) but with the exception that the failed mining equipment is repaired. More specifically, when the mining equipment fails in the field, the repair is attempted. If it cannot be properly repaired, then it is taken to the repair workshop for repair. The redrawn diagram of Figure 11.5 with constant repair rates, μ_1, μ_2, and μ_3 is shown in Figure 11.6. Thus, Figure 11.6 is the state space diagram for this mathematical model.

The assumptions pertaining to this mathematical model are the same as for Model VI. The only additional assumption is that the repaired equipment is as good as new. The equipment repair rates μ_1, μ_2, and μ_3 are defined below:

μ_1 is the constant repair rate of the mining equipment from the unsafely failed state (i.e., state 1).
μ_2 is the constant repair rate of the mining equipment from the safely failed state (i.e., state 2).
μ_3 is the constant repair rate of the mining equipment from the repair workshop (i.e., state 3).

Other symbols used in this mathematical model are exactly the same as for Model VI. Using the Markov method described in Chapter 4 and [1, 3], we write down the following system of differential equations for Figure 11.6:

$$\frac{dP_0(t)}{dt} + (\lambda_s + \lambda_u)P_0(t) = \mu_1 P_1(t) + \mu_2 P_2(t) + \mu_3 P_3(t) \tag{11.50}$$

$$\frac{dP_1(t)}{dt} + (\lambda_1 + \mu_1)P_1(t) = \lambda_u P_0(t) \tag{11.51}$$

$$\frac{dP_2(t)}{dt} + (\lambda_2 + \mu_2)P_2(t) = \lambda_s P_0(t) \tag{11.52}$$

$$\frac{dP_3(t)}{dt} + \mu_3 P_3(t) = \lambda_2 P_2(t) + \lambda_1 P_1(t) \tag{11.53}$$

At time $t=0$, $P_0(0)=1$, and $P_1(0)=P_2(0)=P_3(0)=0$.

By setting the derivatives equal to zero in Equations 11.50–11.53 and utilizing the relationship $\sum_{i=0}^{3} P_i = 1$, we obtain the following steady state probability equations [12]:

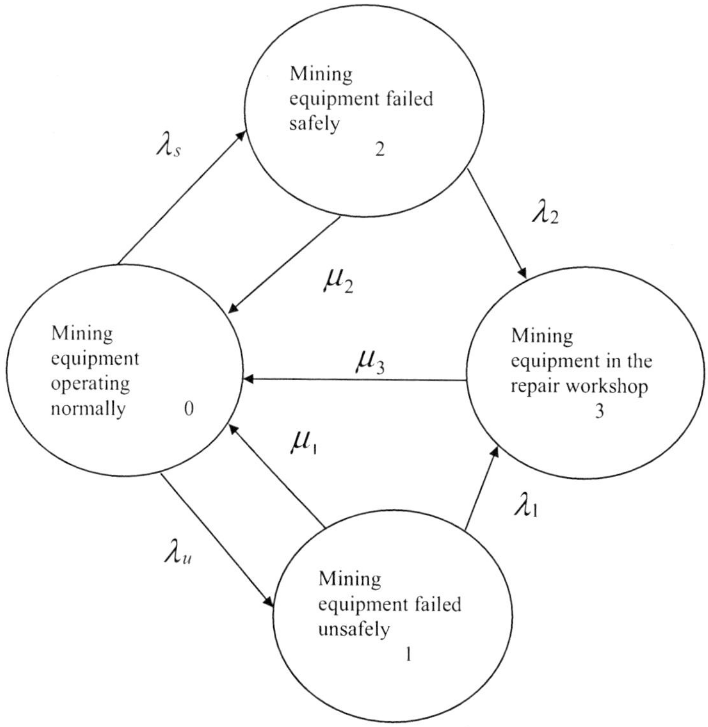

Figure 11.6 Redrawn Figure 11.5 diagram with repair rates

$$P_0 = \frac{D}{E} \tag{11.54}$$

and

$$P_i = n_i P_0, \text{for } i = 1,2,3 \tag{11.55}$$

where

$$D = \mu_3(\lambda_1 + \mu_1)(\lambda_2 + \mu_2)$$

$$E = (\lambda_1 + \mu_1)[\mu_3(\lambda_2 + \mu_2) + \lambda_s(\mu_3 + \lambda_2)] + \lambda_u(\lambda_2 + \mu_2)(\lambda_1 + \mu_3)$$

$$n_1 = \lambda_u / (\lambda_1 + \mu_1)$$

$$n_2 = \lambda_s / (\lambda_2 + \mu_2)$$

$$n_3 = \frac{\lambda_u \lambda_1 (\lambda_2 + \mu_2) + \lambda_s \lambda_2 (\lambda_1 + \mu_1)}{\lambda_3 (\lambda_1 + \mu_1)(\lambda_2 + \mu_2)}$$

P_i is the steady state probability of the mining equipment being in state i, for $i = 0, 1, 2, 3$.

The mining equipment steady state availability and unavailability are given by

$$MESSA = P_0 \tag{11.56}$$

and

$$MESSUA = P_1 + P_2 + P_3 \tag{11.57}$$

where

MESSA is the mining equipment steady state availability.
MESSUA is the mining equipment steady state unavailability.

11.9 Model VIII

This mathematical model represents a mining system that can either fail safely or fail with an accident due to hardware failures or human errors [3, 13]. The failed

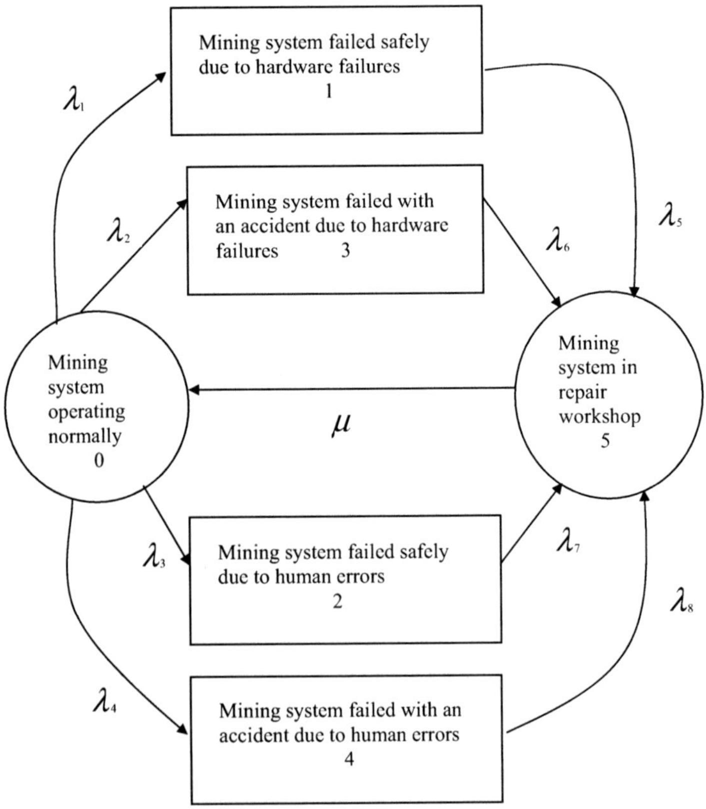

Figure 11.7 State space diagram for the mining system that can either fail safely or fail with an accident due to human errors or hardware failures

mining system is taken to repair workshop for repair. After repair, the mining system is put back into normal operation. The system state space diagram is shown in Figure 11.7. The numerals in boxes and circles denote system states.

The following assumptions are associated with this mathematical model:

- Failures and human errors occur independently.
- Hardware failure rates, human error rates, and rates of taking the failed mining system to the repair workshop are constant.
- The failed mining system repair rate is constant.
- The repaired mining system is as good as new.

The following symbols are associated with this mathematical model:

j	is the jth state of the mining system, where $j = 0$ means that the mining system is operating normally, $j = 1$ means that the mining system has failed safely due to hardware failures, $j = 2$ means that the mining system has failed safely due to human errors, $j = 3$ means that the mining system has failed with an accident due to hardware failures, $j = 4$ means that the mining system has failed with an accident due to human errors, and j=5 means that the mining system is in the repair workshop.
$P_j(t)$	is the probability that the mining system is in state j at time t, for $j = 0$, 1, 2, 3, 4, 5.
λ_1	is the constant hardware failure rate of the mining system failing safely.
λ_2	is the constant hardware failure rate of the mining system that causes an accident.
λ_3	is the constant human error rate of the mining system failing safely.
λ_4	is the constant human error rate of the mining system that causes an accident.
λ_5	is the constant rate of taking the failed mining system to the repair workshop for repair from state 1.
λ_6	is the constant rate of taking the failed mining system to the repair workshop for repair from state 3.
λ_7	is the constant rate of taking the failed mining system to the repair workshop for repair from state 2.
λ_8	is the constant rate of taking the failed mining system to the repair workshop for repair from state 4.
μ	is the constant repair rate of the mining system from state 5 to state 0.
P_j	is the steady state probability that the mining system is in state j, for $j = 0, 1, 2, 3, 4, 5$.

Using the Markov method described in Chapter 4 and [1, 3], we write down the following set of differential equations for Figure 11.7:

$$\frac{dP_0(t)}{dt} + (\lambda_1 + \lambda_2 + \lambda_3 + \lambda_4)P_0(t) = \mu P_5(t) \tag{11.58}$$

$$\frac{dP_1(t)}{dt} + \lambda_5 P_1(t) = \lambda_1 P_0(t) \tag{11.59}$$

$$\frac{dP_2(t)}{dt} + \lambda_7 P_2(t) = \lambda_3 P_0(t) \tag{11.60}$$

$$\frac{dP_3(t)}{dt} + \lambda_6 P_3(t) = \lambda_2 P_0(t) \tag{11.61}$$

$$\frac{dP_4(t)}{dt} + \lambda_8 P_4(t) = \lambda_4 P_0(t) \tag{11.62}$$

$$\frac{dP_5(t)}{dt} + \mu P_5(t) = \lambda_5 P_1(t) + \lambda_6 P_3(t) + \lambda_7 P_2(t) + \lambda_8 P_4(t) \tag{11.63}$$

At time $t=0$, $P_0(0)=1$, and $P_1(0)=P_2(0)=P_3(0)=P_4(0)=P_5(0)=0$.

By setting $\mu=0$ in Equations 11.58–11.64 and then solving for $P_0(t)$, we obtain

$$R_{ms}(t) = P_0(t) = e^{-(\lambda_1+\lambda_2+\lambda_3+\lambda_4)t} \tag{11.64}$$

where

$R_{ms}(t)$ is the mining system reliability at time t.

The mining system mean time to failure ($MTTF_{ms}$) is given by [12]

$$\begin{aligned} MTTF_{ms} &= \int_0^\infty R_{ms}(t)\,dt \\ &= \int_0^\infty e^{-(\lambda_1+\lambda_2+\lambda_3+\lambda_4)t}\,dt \\ &= \frac{1}{(\lambda_1+\lambda_2+\lambda_3+\lambda_4)} \end{aligned} \tag{11.65}$$

By setting derivatives equal to zero in Equations 11.58–11.63 and using the relationship $\sum_{j=0}^{5} P_j = 1$, we get the following set of steady state probability equations [12]:

$$P_0 = \frac{1}{1+M_1} \tag{11.66}$$

where

$$M_1 = \frac{\lambda_1}{\lambda_5} + \frac{\lambda_3}{\lambda_7} + \frac{\lambda_2}{\lambda_6} + \frac{\lambda_4}{\lambda_8} + \frac{M_2}{\mu}$$

where

$$M_2 = \lambda_1 + \lambda_3 + \lambda_2 + \lambda_4$$

$$P_1 = \frac{\lambda_1}{\lambda_5} P_0 \quad (11.67)$$

$$P_2 = \frac{\lambda_3}{\lambda_7} P_0 \quad (11.68)$$

$$P_3 = \frac{\lambda_2}{\lambda_6} P_0 \quad (11.69)$$

$$P_4 = \frac{\lambda_4}{\lambda_8} P_0 \quad (11.70)$$

$$P_5 = \frac{M_2}{\mu} P_0 \quad (11.71)$$

The mining system steady state availability and unavailability are given by

$$AV_{\text{ms}} = P_0 \quad (11.72)$$

$$UAV_{\text{ms}} = \sum_{j=1}^{5} P_j \quad (11.73)$$

where

AV_{ms} is the mining system steady state availability.
UAV_{ms} is the mining system steady state unavailability.

The steady state probability of the mining system failing safely is given by

$$P_{\text{mss}} = P_1 + P_2 \quad (11.74)$$

Similarly, the steady state probability of the mining system with an accident is expressed by

$$P_{\text{msa}} = P_3 + P_4 \quad (11.75)$$

The steady state probability of the mining system failing due to hardware failures is expressed by

$$P_{\text{msh}} = P_1 + P_3 \quad (11.76)$$

Finally, the steady state probability of the mining system failing due to human error is given by

$$P_{\text{mshe}} = P_2 + P_4 \quad (11.77)$$

Example 11.7

Assume that in Figure 11.7 we have the following given values for some transition rates:

$\lambda_1 = 0.0009$ failures/h
$\lambda_2 = 0.0003$ failures/h
$\lambda_3 = 0.0004$ errors/h
$\lambda_4 = 0.0001$ errors/h

Calculate the mining system reliability for a 7-h mission and the mean time to failure.

By substituting the specified data values into Equation 11.64, we get

$$\begin{aligned} R_{ms}(7) &= e^{-(0.0009+0.0003+0.0004+0.0001)(7)} \\ &= 0.9882 \end{aligned}$$

Similarly, by substituting the given data values into Equation 11.65 yields

$$\begin{aligned} MTTF_{ms} &= \frac{1}{(0.0009+0.0003+0.0004+0.0001)} \\ &= 588.23\,\text{h} \end{aligned}$$

Thus, the mining system reliability and mean time to failure for the specified data values are 0.9882 and 588.23 hr, respectively.

11.10 Problems

1. Write an essay on mathematical models used for performing safety analysis in mines.
2. A piece of mining equipment can fail safely or unsafely and its constant failure rates are 0.006 failures per hour and 0.002 failures per hour, respectively. Calculate the probability of the mining equipment failing unsafely during a 50-h mission.
3. Prove Equation 11.20 by using Equation 11.19.
4. Assume that a mining worker is performing a time continuous task and his/her error rate is 0.002 errors per hour. Calculate the worker's reliability for a 10-h mission.
5. A worker is performing a mining task in fluctuating (i.e., normal and stressful) environments and his/her error rates are 0.002 errors/hour and 0.006 errors/hour, respectively. The constant transition rates from normal to stressful environments and *vice versa* are 0.002/h and 0.001/h, respectively. Calculate the mean time to worker error.
6. Prove Equations 11.35–11.37 by using Equations 11.32–11.34.

7. Prove that the sum of Equations 11.44–11.47 is equal to unity.
8. Prove Equations 11.54–11.55 by using Equations 11.50–11.53.
9. Assume that in Figure 11.7, we have the following specified values for some transition rates:

   $\lambda_1 = 0.0008$ failures/hour
   $\lambda_2 = 0.0002$ failures/hour
   $\lambda_3 = 0.0003$ errors/hour
   $\lambda_4 = 0.0001$ errors/hour

   Calculate the mining system reliability for a 9-h mission and the mean time to failure.

10. Prove that the sum of Equations 11.54 and 11.55 is equal to unity.

References

1. Shooman ML. Probabilistic Reliability: An Engineering Approach. New York: McGraw Hill Book Company; 1968.
2. Regulinski TL, Askren WB. Stochastic Modeling of Human Performance Effectiveness Functions. Proceedings of the Annual Reliability and Maintainability Symposium 1972:407–416.
3. Dhillon BS. Human Reliability: with Human Factors. New York: Pergamon Press; 1986.
4. Dhillon BS. Human Reliability and Error in Medical System. River Edge, New York: World Scientific Publishing; 2003.
5. Dhillon BS. Human Reliability, Error, and Human Factors in Engineering Maintenance: with Reference to Aviation and Power Generation. Boca Raton, Florida: CRC Press; 2009.
6. Dhillon BS. Human Reliability and Error in Transportation Systems. London: Springer Inc.; 2007.
7. Dhillon BS. The Analysis of the Reliability of Multi-State Device Networks. Dissertation 1975. Available from the National Library of Canada, Ottawa, Canada.
8. Askren WB, Regulinski TL. Quantifying Human Performance for Reliability Analysis of Systems. Human Factors 1969;11:393–396.
9. Regulinski TL, Askren WB. Mathematical Modeling of Human Performance Reliability. Proceedings of the Annual Symposium on Reliability 1969:5–11.
10. Dhillon BS. Stochastic Models for Predicting Human Reliability. Microelectronics and Reliability 1985;25:729–752.
11. Dhillon BS, Rayapati SN. Reliability and Availability Analysis of Surface Transit Systems. Microelectronics and Reliability 1984;24:1029–1033.
12. Dhillon BS. Design Reliability: Fundamentals and Applications. Boca Raton, Florida: CRC Press; 1999.
13. Dhillon BS, Rayapati SN. Reliability Evaluation of Transportation Systems with Human Errors. Proceedings of the IASTED Int. Conf. on Applied Simulation and Modeling 1985:4–7.

Bibliography

Introduction

Over the years a large number of publications on various aspects of safety in mining have appeared in the form of technical reports, journal articles, conference proceedings articles, etc. This appendix presents a list of over four hundred such publications. The period covered by this listing is from 1865–2009. The main objective of this listing is to provide readers with sources to obtain additional information on safety in mining.

Publications

1. Adhikary DP, Shen B, Duncan ME. A study of highwall mining panel stability. International Journal of Rock Mechanics and Mining Sciences 2002; 39(5): 643–659.
2. Akcil A. Managing cyanide: Health, safety and risk management practices at Turkey's Ovacik gold-silver mine. Journal of Cleaner Production 2006; 14(8): 727–735.
3. Alameddin AN. Industrial coal-fired explosion hazards. Proceedings of the IEEE Cement Industry Technical Conference 1989:225–247.
4. Alois A. Review of the use of nitrogen in mine fires. Transactions of the Institution of Mining and Metallurgy 2008; 111:89–98.
5. Ambrose DH, Bartels JR, Kwitowski AJ, Gallagher S, Battenhouse Jr T. Computer simulations help determine safe vertical boom speeds for roof bolting in underground coal mines. Journal of Safety Research 2005; 36(44):387–397.
6. American Standard Safety Code for Installing and Using Electrical Equipment in and About Coal Mines. Government Printing Office. Washington, D.C.; 1952.
7. Ames RG. Reger RB, Wheeler RW. Injury incidence in underground coal mines. Mining Engineering, Littleton 1983; 35(4):310–312.

8. Amyotte PR, Ochmen AM. Application of a loss causation model to the Westray mine explosion. Process Safety and Environmental Protection 2002; 80:55–59.
9. Amyotte PR, Oehmen AM. Application of a loss causation model to the Westray mine explosion. Process Safety and Environmental Protection 2002; 80(1):55–59.
10. Andrews J. Quantitative safety assessment of the ventilation recirculation system in an undersea mine. Quality and Reliability Engineering International 1991; 7(6):497–510.
11. Andy T. Safety is a bargain at Kidd Creek Mine. Canadian Mining Journal 1984; 105(7):34–35.
12. Ann GK. Locating fires in abandoned underground coal mines. International Journal of Coal Geology 2004; 54(1–2):49–62.
13. Anon. Vanguard of mine safety. Canadian Mining Journal 1996: 2.
14. Anon. MMS safety alerts pinpoint offshore work hazards. Oil and Gas Journal 1983; 81(18):160–172.
15. Anon. Safety in mines research. Iron and Coal Trades Review 1952; 164(4394): 1437–1439.
16. Anon. Five Recent Offshore Fires Produce New Safety Alerts. Oil and Gas Journal 1983; 81(52):154–155.
17. Anon. Kentucky board winks at mine safety. Coal Age 2001; 106(2):22.
18. Anon. Safeguarding the miner. Scientific American 1923; 129(3):174–175.
19. Anon. Safety air in coal mines. The Lancet 1912; 79(4625):1072.
20. Anon. Safety in mines research. Engineer 1930; 150(3892):176–177.
21. Anon. Safety in mines. Canadian Mining Journal 1958; 79(9):81–128.
22. Anon. Safety in mines. Electrical Review 1943; 132(3400):125.
23. Anon. The color of safety. Engineering and Mining Journal 2006; 207(5):46–49.
24. Anon. AEP underground mines safest in U.S.. Coal Age 1997; 102(12):10.
25. Anon. China: Coal mine deaths declined in first half of year-report. Dow Jones International News 2007 July.
26. Anon. Coal mines closed in safety campaign. China Economic Review 2008 Jan 15.
27. Anon. Fire and life safety for underground facilities: Present status of fire and life safety principles related to underground facilities. Tunneling and Underground Space Technology 1998; 13(13):217–269.
28. Anon. Industry, government review mine safety procedures. Mining Engineering 2006; 58(3):45.
29. Anon. Mine "strikes it rich" with self-sufficient fire safety system. Engineered Systems 2002; 19(5):29.
30. Anon. Mine safety. Engineering and Mining Journal 1998; 199(5):32.
31. Anon. MSHA fines Kentucky operator $360,000 after mine fatalities. Coal Age 2006; 111(6):13–14.
32. Anon. New safety light for coal mines. Journal of the Franklin Institute 1865; 79(4):262–263.
33. Anon. New safety mine cage. Canadian Mining Journal 1954; 66(9):612–613.

34. Anon. Rock stress monitor improves mine safety. Canadian Institution of Mining Bulletin 1982; 75(845):179.
35. Anon. Safety in mines. The Lancet 1950; 256(6623):217–218.
36. Anon. Southern minerals fined, ordered to improve safety. Coal Age 1997; 102(2):12.
37. Anon. U.S. Enacts new mine safety policies. Coal Age 2006; 111(6):24–27.
38. Anon. Health and safety in mines. The Lancet 1950; 255(6610): 868.
39. Anon. Health and safety in coal-mines. The Lancet 1939; 233(6019):35–36.
40. Ash SH. Safety Practices in California Gold Dredging. Report. Government Printing Office. Washington D.C.; 1932.
41. Ash SH. Mineral industry health and safety. Mining Engineering 1954; 6(2): 195–196.
42. Ash SH. Mineral industry health and safety. Mining Engineering 1955; 7(3): 283–284.
43. Attwood D, Khan F, Veitch B. Offshore oil and gas occupational accidents-What is important? Journal of Loss Prevention in the Process Industries 2006; 19(5):386–398.
44. Attwood D, Khan F, Veitch B. Can we predict occupational accident frequency? Process Safety and Environmental Protection 2006; 84(3):208–221.
45. Bajpayee TS, Rehak TR, Ingram D.K. Blasting injuries in surface mining with emphasis on flyrock and blast area security. Journal of Safety Research 2004; 35(1):47–57.
46. Baker SP, Conroy C, Johnston JJ. Occupational injury prevention. Journal of Safety Research 1992; 23(2):129–133.
47. Banerjee BD, Dhar BB. Issues and options for reducing methane emission to the atmosphere from Indian coal mining Energy Conversion and Management 1996; 37(6–8):1175–1179.
48. Barkand TD. Safe electrical design of mine elevator control systems. IEEE Transactions on Industry Applications 1997; 33(2):312–318.
49. Barkand TD. Ascending elevator accidents: Give the miner a brake. IEEE Transactions on Industry Applications 1992; 28(3):720–729.
50. Barkand TD, Helfrich W J. Application of dynamic braking to mine hoisting systems. IEEE Transactions on Industry Applications 1988; 24(5):884–896.
51. Bauer ER, Bise CJ. Deep cut: Ground control and worker safety in coal mines. Mining Engineering, Littleton 2000; 52(6):110–116.
52. Bauer ER, Chekan GJ, Steiner LJ. Stability evaluation of extended cut mining in underground coal mines. International Journal of Rock Mechanics and Mining Sciences 1997; 34(3–4):302.
53. Bea RG. Human and organization factors: engineering operating safety into offshore structures. Reliability Engineering & System Safety 1998; 61(1–2): 109–126.
54. Beamish BB, McKavanagh BM. Underground Rock-Noise Monitoring At No. 2 Mine, Collinsville. Geoexploration 1987; 24(4–5):347–365.
55. Beamish BB, Crosdale JP. Instantaneous outbursts in underground coal mines: an overview and association with coal type. International Journal of Coal Geology 1987; 35:27–55.

56. Bellamy LJ. Influence of human factors science on safety in the offshore industry. Journal of Loss Prevention in the Process Industries 1994; 7(4):370–375.
57. Bennett JD, Passmore DL. Correlates of coal mine accidents and injuries: A literature Review. Accident Analysis and Prevention 1984; 16(1):37–45.
58. Bennett JD, Passmore DL. Days lost from work due to injuries in the US underground bituminous coal mines. Journal of Occupational Accidents 1984; 5:265–278.
59. Beus MJ, Ruff TM, McCoy WG. Conveyance monitoring to improve mine hoisting safety. Proceeding of the 32nd IEEE Industry Applications Society Annual Meeting 1997:2091–2097.
60. Bhattacherjee A, Ramani RV, Natarajan R. Time series analysis of coal mine accident experience. Journal of Safety Research 1994; 25(4):229–234.
61. Bibliography of Bureau of Mines Publications Dealing with Health and Safety in the Mineral and Allied Industries. Report, Bureau of Mines. United States Dept. of the Interior. Washington D.C.; 1946.
62. Bigourd J, Michot C. Explosives research at CERCHAR. Propellants, Explosives, Pyrotechnics 1989; 14(3):103–107.
63. Bise CJ. Non-intrusive safety-and-training sample strategy for the continuous miner worksite. Mining Engineering, Littleton 1990; 42(8):1002–1006.
64. Bise CJ, Peck KL. Microcomputer-based instruction for miner safety-training program. Computers & Industrial Engineering 1991; 98:413–417.
65. Blignaut CJH. The perception of Hazard: Hazard analysis and the contribution of visual search to hazard perception. Ergonomics 1979; 22:991–999.
66. Boardman JL. What Should a Mine Safety Inspector Look for? Canadian Mining Journal 1941; 62(6):386–389.
67. Born G, Giese R, Otto P, Dickmann TH, Amberg F. Integrated seismic imaging system for the geologic prediction ahead in underground construction. Proceedings of the Society of Mining Engineers Rapid Excavation and Tunneling Conference 2001:263–271.
68. Borse E. Design basis accidents and accident analysis with particular reference to offshore platforms. Journal of Occupational Accidents 1979; 2(3):227–243.
69. Bowers BA. Mine safety to 1999. Canadian Institution Mining Bulletin 1974; 67(751):112–117.
70. Bowles O. Safety in Stone Quarrying. Government Printing Office. Washington, D.C.; 1915.
71. Breuer J, Hoffer EM, Hummitzsch W. Rate of occupational accidents in the mining industry since 1950 – A successful approach to prevention policy. Journal of Safety Research 2002; 33(1):129–141.
72. Brown D, Stile JK, Gunderson L, Giras TC. Mining human failure dynamics from accident data using logistic regression and decision trees. Proceedings of the IEEE International Conference on Systems 2002:4–6.
73. Brunton DW, Davis JA. Safety and Efficiency in Mine Tunneling. Government Printing Office. Washington D.C.; 1916.
74. Bryan A. Safety in mines. Engineering 1951; 171(4451):608–610.
75. Bryan AM. The evaluation of Health and Safety in Mines. London: Ashire Publishing; 1975.

76. Burgess-Limerick R, Straker L, Pollock C, Dennis G, Leveritt S, Johnson S. Implementation of the participative ergonomics for manual tasks programme at four Australian underground coal mines. International Journal of Industrial Ergonomics 2007; 37(2):145–155.
77. Burns D. Safety in Coal Mines: A Textbook of Fundamental Principles for Firemen and other Workers in Mines. New York: D.Van Nostrand; 1912.
78. Butani SJ. Relative risk analysis of injuries in coal mining by age and experience at present company. Journal of Occupational Accidents 1988; 10(3): 209–216.
79. Butler HR, Thomson B, McLeod PC, Taylor BW, Froham A, Arnew RS, Greenfield MS, Cummings T, Hinds DG, Weglo JK, Chakravati JL, Hay WR, Falconer D, Lynch D. Do's and don'ts on hazardous materials. Canadian Mining Journal 1973; 94(9):23–26, 28, 30, 32.
80. Camm T, Girard-Dwyer J. Economic consequences of mining injuries. Mining Engineering 2005; 57(9):89–92.
81. Campbell AD. Place of management in safety. Canadian Mining Journal 1941; 6(1):29–31.
82. Cao Y, He D, Glick DC. Coal and gas outburst in footwalls of reverse faults. International Journal of Coal Geology 2001; 48(1–2):47–63.
83. Cashdollar KL, Hertzberg M. Laboratory study of rock dust inerting requirements. Effects of coal volatility, particle size and methane addition. Proceedings of the 23rd international conference on safety in mines 1989: 965–977.
84. Cawley JC. Electrical accidents in the mining industry, 1990–1999. IEEE Transactions on Industry Applications 2003; 39(6):1570–1577.
85. Cawley JC. A statistical determination of spark ignition safety factors in Methane, Propane, and Ethylene mixtures in air. Bureau of Mines. United States Dept. of the Interior, Pittsburgh, Pennsylvania; 1986.
86. Chajet H. New risks from the new MINER Law. Coal Age 2006; 111(6):48.
87. Chandler RJ, Tosatti G. Stava tailings dams failure. Geotechnical Engineering 1995; 113(2):67–79.
88. Chang W, McMechan GA. Reverse-time migration of offset vertical seismic profiling data using the excitation-time imaging condition. Geophysics 1986; 51:67–84.
89. Chen F. Smoke propagation in road tunnels. ASME Applied Mechanics 2000; 53(8):207–218.
90. Chen SG, Zhao J. Use of monitoring records and management system in Ertan cavern construction. Rock Mechanics and Rock Engineering 1998; 31(2):125–131.
91. Cholewa W. Operational safety of hoisting ropes. Mining Science & Technology 1986; 3(3):181–186.
92. Chouteau M, Krivochieva S, Carvalho A. Carter JE, Singhai R. Application of in-mine time-domain electromagnetic methods for mine planning and safety at Taquari-Vassouras potash mine (Brazil). Proceedings of the Canadian Institute of Mining 101st Annual General Meeting 1999:109–110.

93. Clanzy TK. Safety in mines-past, present and future or can we profit from experience? Mining Technology 1979; 61(706):369–381.
94. Clark HH. Suggested Safety Rules for Installing and Using Electrical Equipment in Bituminous Coal Mines. Report. Government Printing Office. Washington D.C.; 1916.
95. Clive N. Studies of accidents leading to minor injuries in the U.K. coal mining industry. Journal of Occupational Accidents 1980; 2(4):305–323.
96. Coal Mine Safety Inspection Manual for Underground Mines. Report. United States Bureau of Mines. Washington, D.C.; 1971.
97. Coleman PJ, Kerkering JC. Measuring mining safety with injury statistics: Lost workdays as indicators of risk. Journal of Safety Research 2007; 38(5): 523–533.
98. Collins EW. Safety evaluation of coal mine power systems. Proceedings of the Annual Reliability and Maintainability Symposium, 1987; 51–56.
99. Collins L. Close call in trench/excavation rescue. Fire Engineering 2004; 157(9):51–62.
100. Conchie SM, Donald IJ. The role of distrust in offshore safety performance. Risk Analysis 2006; 26(5):1151–1159.
101. Cooley WL, Hill HW. Mine power-system grounding research. IEEE Transactions on Industry Applications 1988; 24(5):846–852.
102. Cox SJ, Cheyne AJT. Assessing safety culture in offshore environments. Safety Science 2000; 34(1–3):111–129.
103. Crawley FK. Change in safety management for offshore oil and gas production systems. Process Safety and Environmental Protection 1999; 77(3): 143–148.
104. Crichton M. Attitudes to teamwork, leadership, and stress in oil industry drilling teams. Safety Science 2005; 43(9):679–696.
105. Cross J, Walters M. Vibration and jarring as a cause of back injury in the NSW coal mining industry. Safety Science 1994; 17(4):269–274.
106. Dalvenry LE, Cohen KK, Mowery GL, Kim AG. The use of multiple geophysical techniques for site assessment in mine reclamation. Proceedings of the Symposium on Application of Geophysics to Engineering and Environmental Problems 1996: 887–896.
107. Dalvenry LE, Kim AG. Magnetometer and terrain conductivity surveys at a burning coal west bank. Proceedings of the 17th Annual Abandoned Mine Land Conference 1995:49–69.
108. David L. Safety rules and regulations on mine sites – The problem and a solution. Journal of Safety Research 2005; 36(1):39–50.
109. deMeulles R. Measuring safety: How knowledge transfer supported an industry-wide safety initiative in Ontario mines. Canadian of Institution Mining Bulletin 2002; 95(1066):65–66.
110. Denby B, Schofield D. Role of virtual reality in safety training of mine personnel. Mining Engineering 1999; 51(10):59–64.
111. Denton AA. Safety offshore – loss prevention in a pioneering industry. Proceedings of the Institution of Mechanical Engineers. Journal of Process Mechanical Engineering 1993; 207(2):79–97.

112. Denzler HE. How safety is designed into offshore platforms. World Oil 1961; 152(7):131–133, 136.
113. Dhananjay K, Bengt K, Uday K. Reliability analysis of power transmission cables of electric mine loaders using the proportional hazards model. Reliability Engineering & System Safety 992; 37(3):217–222.
114. Dhillon BS. Mining Equipment Reliability, Maintainability, and Safety. London: Springer Inc; 2008.
115. Diaz A, Maria B, Gonzalez NC. Control and prevention of gas outbursts in coal mines. Journal of Coal Geology 2007:253–266.
116. Dino GM. Lower oil prices, safety influence offshore design trends. Oil and Gas Journal 1989; 87(32):71–72, 74–76.
117. Dodge CE. Educating the miner in safety. Canadian Mining Journal 1928; 49(39):786–788.
118. Douglass DP. Safety devices on electric hoists used in Ontario mines. Canadian Mining Journal 1940; 61(4):229–234.
119. Duchon JC, Smith TJ. Extended workdays and safety. International Journal of Industrial Ergonomics 1993; 11(1):37–49.
120. Duzgun HSB, Einstein HH. Assessment and management of roof fall risks in underground coal mines. Safety Science 2004; 42(1):23–41.
121. Eisner HS, Leger JP. International safety rating system in south African mining. Journal of Occupational Accidents 1988; 10(2):141–160.
122. Eisner HS, Leger JP. The international safety rating system in South African mining. Journal of Occupational Accidents 1988; 10(2):141–160.
123. Ellen S, Melanie A. Texas sand company pays MSHA $20,000 to settle fatality case. Pit and Quarry 2003; 95(7):53.
124. Fadeev AB, Glosman LV, Sofonov LV. Seismic control of mine and quarry blasting in the USSR. Proceedings of the International Congress on Rock Mechanics 1987: 617–619.
125. Falker III JM, Nickerson W. A new direction for safety policy: The offshore oil industry and safety regulation of technology. Technology in Society 1996; 18(4):503–510.
126. Fatal Accidents Involving Roof Falls in Coal Mining, 1996–1998. Report. Mine Safety and Health Administration. United States Dept. of Labor. Washington D.C.; 1998.
127. Fetta B, Mallet LG. Effects of Mining Height on Injury Rates in U.S. Underground Nonlongwall Bituminous Coal Mines. National Institute for Occupational Safety and Health. United States Dept. of Health and Human Services. Pittsburgh, Pennsylvania; 1997.
128. Fiscor S. Miner act will reshape the underground coal mining business. Coal Age 2006; 111(6):2.
129. Flores RM. Coal bed methane: From hazard to resource. International Journal of Coal Geology 1998; 35(1–4):3–26.
130. Forbes JJ. Federal coal mine safety act. Mining Engineering 1954; 6(1):76–78.
131. Francois H, Jackie W. A mining case study: The safe maintenance of underground railway track. Civil Engineering 2006; 14(5):8–10.

132. Franks JJ, Bransford JD. Abstraction of visual patterns. Journal of Experimental Psychology 1971; 90:65–74.
133. Galvin J. Health and safety in Australia's mines. Materials World 2006; 14(4):22–23.
134. Gao PZ, Liu SL, Chow WK, Fong NK. Large eddy simulations for studying tunnel smoke ventilation. Tunneling and Underground Space Technology 2004; 19(6):577–586.
135. Gardner LU. Dust. Canadian Mining Journal 1934; 55(8):370–371.
136. Gary JC. Prevention, control and/or extinguishment of coal seam fires using cellular grout. International Journal of Coal Geology 2004; 59(1–2):75–81.
137. Gedge G. Safe practices underground at Hollinger. Canadian Mining Journal 1943; 64(3):142–146.
138. Gedge G. Good housekeeping as aid to safety. Canadian Mining Journal 1944; 65(8): 549–552.
139. Gerald SP, Hugh BM, James R, Carmel MB, Margaret K, Robin BH, Jefferey LB. International evaluation of injury rates in coal mining: A comparison of risk and compliance-based regulatory approaches. Safety Science 2008; 46(8):1196–1204.
140. Ghosh AK, Bhattacherjee A. Predictors of occupational injuries among coal miners: Causal analysis. Transactions of the Institutions of Mining and Metallurgy 2007; 116(1):16–24.
141. Giardino DA. Estimating the rate of coal combustion in a mine fire. Mine Safety and Health Administration. United States Dept. of Labor. Washington D.C.; 1999.
142. Gibson CS. Ventilation as means to safety. Canadian Mining Journal 1940; 61(12):814–816.
143. Gibson CS. Aspects of safety affecting mechanical and electrical services in mines. Canadian Mining Journal 1952; 73(5):66–71.
144. Gipps GW. Identification and quantification of health and safety risks in mining. Canadian Institution of Mining Bulletin 1981; 74(827):164–166.
145. Gleick PH. Health and safety effects of coal transportation: Reassessing the risks. Energy 1981; 6(7):611–619.
146. Gochioco LM, Rodriguez R, Ruev F. Employing the underground inseam seismic method to detect old abandoned mines and improve mine safety. Leading Edge 2008; 27 (1):88–97.
147. Gochioco LM, Miller T, Ruev F. High-resolution 2D surface seismic reflection survey to detect abandoned old coal mine works to improve mine safety. Leading Edge 2008; 27(1):80–86.
148. Gochioco LM. Coal, geophysics, U.S. power generation, and mine safety. Leading Edge 2004; 23(11):1132–1134.
149. Gonzalez-Nicieza C, Alvarez-Fernandez MI, Menendez-Diaz A, Alvarez-Vigil AE. Comparative analysis of pillar design methods and its application to marble mines. Rock Mechanics and Rock Engineering 2006; 39(5):421–444.
150. Gonzalez-Nicieza C. et al. A comparative analysis of pillar design methods and its application to marble mines. Rock Mechanics and Rock Engineering 2006; 39(5):421–444.

151. Goraya A, Amyotte PR, Khan FI. An inherent safety-based incident investigation methodology. Process Safety Progress 2004; 23(3):197–205.
152. Gordon RPE. Contribution of human factors to accidents in the offshore oil industry. Reliability Engineering and System Safety 1998; 61(1–2):95–108.
153. Gordon RL, Manula CB, Fettig C, Gresham JB. Simulating the effects of the Coal Mine Health and Safety Act. Resources and Energy 1981; 3(1):55–63.
154. Gorzalczynski S. Limitations of safety arrest mechanisms for mine shaft conveyances. Canadian Institution of Mining Bulletin 2002; 95(1065):67–71.
155. Gratt LB. Risk analysis of hazardous materials in oil shale. Journal of Hazardous Materials 1985; 10(2–3):317–350.
156. Gray FW. Safety rules for coal mines. Canadian Mining Journal 1950; 71(3): 74–75.
157. Grayson R.L. Safety vs. productivity and other factors in US underground coal mines. Mining Engineering 2001; 53(8):41–44.
158. Grenier MG, Hardcastle SG. Safety and health in the Canadian mine environment: An overview of ventilation, dust, radiation and diesel emission research. Canadian Institute of Mining Bulletin 1990; 83(944):41–46.
159. Griffith FE, Van der Veer HJ. Use of reflector Buttons for danger, warning, direction, and safety signs in mines. Canadian Mining Journal 1938; 59(7): 379–380.
160. Groombridge JC. A coal-mine 'safety case': Suggestions from the petranalysisoleum industry following the Piper Alpha disaster. Transactions of the Institution of Mining and Metallurgy 2001; 64:18–26.
161. Groves WA, Kecojevic VJ, Komljenovic D. Analysis of fatalities and injuries involving mining equipment. Journal of Safety Research 2007; 38(4): 461–470.
162. Gunningham N. Mine Safety: Law Regulation Policy. The Federation Press, Annandale, New South Wales, Australia: 2007.
163. Gyekye AS. Causal attributions of Ghanaian industrial workers for accident occurrence: Miners and non-miners perspective. Journal of Safety Research 2003; 34(5):533–538.
164. Haaverstad TA. Structural response to accidental explosions and fires on offshore process installations. Journal of Loss Prevention in the Process Industries 1994; 7(4):310–316.
165. Haibo C, Torgeir M, Harry V. Safety of dynamic positioning operations on mobile offshore drilling units. Reliability Engineering and System Safety 2008; 5(7):1072–1090.
166. Hanahen PM. Explosives accidents on geophysical crews. Geophysics 1948; 13(3):421–427.
167. Hanson DR, Vandergrift TL, DeMarco MJ, Hanna K. Advanced techniques in site characterization and mining hazard detection for the underground coal industry. International Journal of Coal Geology 2002; 50(1–4):275–301.
168. Harpalani S, Chen G. Estimation of changes in fracture porosity of coal with gas emission. Fuel 1995; 74:1491–1496.

169. Harrington D. Flame safety lamp proves source of danger. Coal Age 1924; 26(18):610–611.
170. Harrington JH, East JH, Warncke RG. Safety in the mining industry. Report. Government Printing Office. Washington. D.C.; 1950.
171. Harris AG. Mine subsidence problems in North Canton, Stark County, Ohio. The Ohio Journal of Science 1997; 97(2):39.
172. Harteis SP, Dolinar DR. Water and slurry bulkheads in underground coal mines; design, monitoring and safety concerns. Mining Engineering 2006; 58,(12):41–47.
173. Helander MG, Krohn GS, Curtin R. Safety of roof-bolting operations in underground coal mines. Journal of Occupational Accidents 1983; 5(3):161–175.
174. Henderson L. Safety engineering at Alabama coal mines. Mining Engineering 1957; 9(8):890–893.
175. Hickman JS, Geller ES. A safety self-management intervention for mining operations. Journal of Safety Research 2003; 34(3):299–308.
176. Hine DW, Lewko J, Blanco J. Alignment to workplace safety principles: An application to mining. Journal of Safety Research 1999; 30(3):173–185.
177. Hokstad P, Oien K, Reinertsen R, Recommendations on the use of expert judgment in safety and reliability engineering studies: Two offshore case studies. Reliability Engineering & System Safety 1998; 61(1–2):65–76.
178. Homce GT, Cawley JC, Yenchek MR, Sacks HK. An alarm to warm of overhead power line contact by mobile equipment. Proceedings of the IEEE Industry Applications Society Conference. 2001; 2:1376–1383.
179. Hopey D. Coal Miner Death Rate Spiking. Pittsburgh Post-Gazette 2006 May 28.
180. Hopkins R. Reducing high frequency ground currents to zero enhances mine safety. Coal Age 2006; 111(11):36–39.
181. Hou YB, Wang Y, Gao Y, Du JY, Wang M, Zhang HF, Jiao SL. The study of fault location and safety control for the mine ventilating fan. Proceedings of the International Conference on Machine Learning and Cybernetics 2006: 3632–3636.
182. Hovden J, Lie T, Karlsen JE, Alteren B. The safety representative under pressure: A study of occupational health and safety management in the Norwegian oil and gas industry. Safety Science 2008; 46(3):493–509.
183. Hsiao H, Simeonov P. Preventing falls from roofs: A critical review. Ergonomics 2001; 44(5):537–561.
184. Hu LH, Huo R, Li YZ, Wang HB, Chow WK. Full-scale burning tests on studying smoke temperature and velocity along a corridor. Tunneling and Underground Space Technology 2005; 20(3):223–229.
185. Hu Y, McMechan GA. Imaging mining hazards within coal beds using prestack wave equation migration of in-seam seismic survey data: A feasibility study with synthetic data. Journal of Applied Geophysics 2007; 74(1):24–34.
186. Hull BP, Leigh J, Driscoll TR, Mandryk J. Factors associated with occupational injury severity in the New South Wales underground coal mining industry. Safety Science 1996; 21(3):191–204.

187. Iannacchione A, Prosser I, Esterhuizen G, Bajpayee T. Methods for determining roof fall risk in underground mines. Mining Engineering 2007; 59(11):47–53.
188. Iannacchione AT, Marshall TE, Burke L, Melville R, Litsenberger J. Safer mine layouts for underground stone mines subjected to excessive levels of horizontal stress. Mining Engineering 2003; 55(4):25–31.
189. Imada A. Participatory ergonomics: a strategy for creating human-centered work. Journal of Science and Labor 2000; 76:25–31.
190. Iskander WH, Nutter RS. Methodology development for safety and reliability analysis for electrical mine monitoring systems. Microelectronics and Reliability 1988; 28(4):581–597.
191. Jincai Z, Baohong S. Coal mining under aquifers in China: A case study. International Journal of Rock Mechanics and Mining Sciences 2004; 41(4): 629–639.
192. Joel K, Duncan S. A practical approach to fire hazard analysis for offshore structures. Journal of Hazardous Materials 2003; 104(1–3):107–122.
193. Jones JH, Trickett JC. Some observations on the examinations of gases resulting from explosions in collieries. Transactions of the Institution of Mining Engineering 1955; 114:768–791.
194. Jorge P, Jose R, Juan D. Safety, reliability and economics in mining systems. Proceedings of the 41st IEEE Industry Applications Society Annual Conference 2006; 2:942–946.
195. Joy J. Occupational safety risk management in Australian mining. Occupation and Medicine 2001; 54:311–315.
196. Kalia HN. Understanding coal geology can improve underground mine productivity and safety. Mining Engineering 1975; 27(12):68.
197. Kallunki G. Safety in pit operations at Munro mine. Canadian Mining Journal 1953; 74(9):103–107.
198. Karacan CÖ, Ulery JP, Goodman GVR. A numerical evaluation on the effects of impermeable faults on degasification efficiency and methane emissions during underground coal mining. International Journal of Coal Geology 2008; 75(4):195–203.
199. Karlinski J, Rusinski E, Smolnicki T. Protective structures for construction and mining machine operators. Automation in Construction 2008; 17(3): 232–244.
200. Karmis M, editor. Mine Health and Safety Management. Society for Mining, Metallurgy, and Exploration. Littleton, Colorado; 2001.
201. Karra VK. Analysis of non-fatal and fatal injury rates for mine operator and contractor employees and the influence of work location. Journal of Safety Research 2005; 36(5):413–421.
202. Kathleen KT, Edward B. Reducing non-contact electric arc injuries: An investigation of behavioral and organizational issues. Journal of Safety Research 2007; 38(5):597–608.
203. Kathryn M, Torbjorn R, Rhona F, Rachael G, Mark F. Evaluation of psychosocial and organizational factors in offshore safety: A comparative study. Journal of Risk Research 2004; 7(5):545–561.

204. Kecojevic V, Radomsky M, The causes and control of loader and truck related fatalities in surface mining operations. Injury Control and Safety Promotion 2004; 11(1):239–251.
205. Kecojevic V, Willis D, Wilkinson W, Schissler A. Computer mapping of faults in coal mining, International Journal of Coal Geology. Geologic Hazards in Coal Mining 2005; 64(1–2):79–84.
206. Kendriok FW. Safeguarding mine buildings against fire. Canadian Mining Journal 1938; 59(1):22–25.
207. Khair AW, Quinn MK, Chaffins RD. Effects of topography on ground movement due to longwall mining. Mining Engineering 1988; 40(8):820–822.
208. Khan FI, Amyotte PR, DiMattia DG. HEPI: A new tool for human error probability calculation for offshore operation. Safety Science 2006; 44(4): 313–334.
209. Khan FI, Amyotte PR. Inherent safety in offshore oil and gas activities: A review of the present status and future directions. Journal of Loss Prevention in the Process Industries 2002; 15(4):279–289.
210. Kim JA, Jung W. A taxonomy of performance influencing factors for human reliability analysis of emergency tasks. Journal of Loss Prevention in the Process Industries 2003; 16:479–495.
211. King JL. Approach to developing a ground-motion design basis for facilities important to safety at Yucca Mountain. Proceeding of the 1st International Topical Meeting on High Level Radioactive Waste Management Part 1 1990:103–108.
212. Kitamura O. Comparative study on collision resistance of side structure. Marine Technology 1997; 34(4):293–308.
213. Kletz TA. Inherently safer plants. Plant/Operation Progress 1985; 4:164–166.
214. Kloos G, Guivant JE, Nebot EM, Masson F. Range based localisation using RF and the application to mining safety. Proceedings of the IEEE International Conference on Intelligent Robots and Systems 2006:1304–1311.
215. Kneib G, Kaseel A, Lorenz K. Automatic seismic prediction ahead of the tunnel bore machine. First Break 2000; 14:295–302.
216. Kock A, Oberholzer JW. The development and application of electronic technology to increase health, safety and productivity in the South African coal mining industry. Proceedings of the 13th IEEE Industry Applications Society Annual Meeting 1995:2017–2022.
217. Kogi K. Job content and working time: The scope of joint change. Ergonomics 1991; 34(6):757–773.
218. Komljenovic D, Groves WA, Kecojevic VJ. Injuries in U.S. mining operations – A preliminary risk analysis. Safety Science 2008; 46(5):792–801.
219. Komljenovic D. Development of risk-informed, performance-based asset management in mining. International Journal of Mining 2008; 22(2):146–153.
220. Kormendi A. Seismic measurements for safety in mines. Geophysical Prospecting 1986; 34(7):1022–1037.
221. Kovalchik PG, Matetic RJ, Smith AK, Bealko SB. Application of prevention through design for hearing loss in the mining industry. Journal of Safety Research 2008; 39(2):251–254.

222. Kowalski-Trakofler KM, Barrett EA. The concept of degraded images applied to hazard recognition training in mining for reduction of lost-time injuries. Journal of Safety Research 2003; 34(5):515–525.
223. Kral S. Mine safety – Always an important item. Mining Engineering 2001; 53(12):36–40.
224. Kristin SF, Roger C. Ethics and choosing appropriate means to an end: Problems with coal mine and nuclear workplace safety. Risk Analysis 2004; 24(1):147–156.
225. Kuenzi JK, Nelson BC. Mobile Mine Equipment Maintenance Safety: A Review of U.S. Bureau of Mines Research. Bureau of Mines. United States Depart of the Interior. Washington D.C.; 1995.
226. Kumral M. Reliability-based optimisation of a mine production system using genetic algorithms. Journal of Loss Prevention in the Process Industries 2005; 18(3):186–189.
227. Kwitowski AJ, Brautigam AL, Monaghan WD. Teleoperated continuous mining machine for improved safety. Mining Engineering 1995; 47(8):753–759.
228. Lama RD, Bodziony J. Management of outburst in underground coal mines. International Journal of Coal Geology 1998; 35:83–115.
229. Lauriski DD, Guymon RM. Safety management: what it means to us. Mining Engineering 1989; 41(10):1032–1035.
230. Lebecki K, Li J, Dyduch Z, Cybulski K. The possible use of solar panels as flame detectors of triggered barriers. Journal of Loss Prevention in the Process Industries 2000; 13(3-5):311–317.
231. Ledford P. Blast kills five coal miners in Kentucky. Coal Age 2006; 111(6):9.
232. Lee BS, Venkataramanan S. Knowledge-based systems approach for offshore safety training. Computers in Industry 1991; 17(4):349–358.
233. Leger JP. Trends and causes of fatalities in South African mines. Safety Science 1991; 14(3-4):169–185.
234. Lei T, Ri-jia D. Efficiency assessment of coal mine safety input by data envelopment analysis. Journal of China University of Mining and Technology 2008; 18(1):88–92.
235. Leigh J, Mulder HB, Want GV, Farnsworth NP , Morgan GG. Sprain/strain back injuries in New South Wales underground coal mining. Safety Science 1991; 14(1): 35–42.
236. Leigh J, Waehrer G, Miller T, Keenan C. Cost of occupational injury and illness across industries. Scandinavian Journal of Work Environment and Health 2004; 30(3):199–205.
237. Li JSM, Chow WK. Illegal carriage of dangerous goods and their effects on tunnel safety. Tunneling and Underground Space Technology 2000; 15(2): 167–173.
238. Li W, Mei S, Zai S, Zhao S, Liang X. Fuzzy models for analysis of rock mass displacements due to underground mining in mountainous areas. International Journal of Rock Mechanics and Mining Sciences 2006; 43(4):503–511.

239. Li W, Mei S. Fuzzy mathematical method of prediction and analysis on rock mass displacements and instability due to underground mining under river valley. China Journal of Rock Mechanical Engineering 2003; 1(20):2289–2293.
240. Lind GH. Risk management of underground coal pillar extraction in South Africa. International Journal of Surface Mining, Reclamation and Environment 2005; 19(3):218–233.
241. Litchfield. et al. Practical Ignition Problems Related to Intrinsic Safety in Mine Equipment: Four Short-term Studies. Report. Bureau of Mines. United States Dept. of the Interior. Washington. D.C.; 1980.
242. Liu YJ, Mao SJ, Li M, Yao JM. Study of a comprehensive assessment method for coal mine safety based on a hierarchical grey analysis. Journal of China University of Mining and Technology 2007; 17(1):6–10.
243. Lucier AM, Heesakkers V, Zoback MD, Reches Z. Estimation of in situ stress magnitudes and orientations in a deep South African gold mine; applications to fault mechanics and mine safety. Proceedings of the American Geophysical Union 87th meeting 2006:10–14.
244. Lutz WK. Advancing inherent safety into methodology. Process Safety Progress 1997; 16(2):86–88.
245. Lyon E. Miners' electric safety lamps. Electrical Review 1926; 98(2510):9–10.
246. Main BW, McMurphy KJ. Safer by design: reducing hazards through better designs. Journal of the American Society of Safety Engineers 1998; 43(2): 29–33.
247. Mainardi S. Earnings and work accident risk: A panel data analysis on mining. Resources Policy 2005; 30(3)156–167.
248. Maiti J, Bhattacherjee A. Evaluation of risk of occupational injuries among underground coal mine workers through multinomial logit analysis. Journal of Safety Research 1999; 30(2):93–101.
249. Mallick S, Mukherjee K. An empirical study for mines safety management through analysis on potential for accident reduction. Omega 1996; 24(5): 539–550.
250. Malone TF. Weather hazards in offshore operations. World Oil 1956; 142(7): 176, 178, 180, 184.
251. Marras WS, Bobick TG, Lavender SA, Rockwell TH, Lundquist, RL. Risks of hand tool injury in U.S. underground mining from 1978 through 1983. Part I: Coal Mining. Journal of Safety Research 1988; 19(2):71–85.
252. Marras WS, Lavender SA, Bobick TG, Rockwell TH, Lundquist RL. Risks of hand tool injury in U.S. underground mining from 1978 through 1983. Part II: metal-nonmetal mining. Journal of Safety Research 1988; 19(3):115–124.
253. Marshall E. Offshore drilling safety questioned. Science 1981; 213(4505): 314–315.
254. Marshall KL. Mine safety as affected by electrification. Coal Mining 1928; 5(3):79–80.
255. Martin LA, Girard JM, Goris JM. Installation and safety practices for cable bolts in underground mines. Bureau of Mines. United States Dept. of the Interior, Pittsburgh, Pennsylvania; 1996.

256. Matzgar C. Jumping to painful conclusions. Pit and Quarry 2003; 96(5):14, 16, 18.
257. McIvor RA. Mine shaft conveyance safety mechanism – free-fall testing. Canadian Mining Journal Bulletin 1996; 89(1004):47–50.
258. McKenna FP. Accident Proneness: A conceptual Analysis. Accident Analysis and Prevention 1983; 15(1):65–71.
259. Mearns K, Whitaker SM, Flin R. Safety climate, safety management practice and safety performance in offshore environments. Safety Science 2003; 41(8):641–680.
260. Megredy HD. Mining safety the North Carolina way. Proceeding of the Ceramic Engineering and Science 1997; 18(2):304–307.
261. Metzgar C. Something for nothing – Almost. Pit and Quarry 2003; 96(2):60.
262. Metzgar C. Trip report II. Pit and Quarry 2003; 95(7):50.
263. Michael KL. Early mining safety heroes. Mining Engineering 2006; 58(3): 46.
264. Mikkelsen A, Ringstad AJ, Steineke, JM. Working time arrangements and safety for offshore workers in the North Sea. Safety Science 2004; 42(3): 167–184.
265. Miller TR. Estimating the cost of injury to US employers. Journal of Safety Research 1997; 28(1):1–13.
266. Mitchell RJ, Driscoll TR, Harrison JE. Traumatic work-related fatalities involving mining in Australia. Safety Science 1998; 29(2):107–123.
267. Miura K, Morooka CK, Mendes JRP, Guilherme IR. Characterization of operational safety in offshore oil wells. Journal of Petroleum Science and Engineering 2006; 51(1–2):111–126.
268. Modic J. Fire simulation in road-tunnels. Tunneling and Underground Space Technology 2003; 18(5):525–530.
269. Molinda GM. Geologic hazards and roof stability in coal mines. Report. National Institute for Occupational Safety and Health. United States Dept. of Health and Human Services. Pittsburgh, Pennsylvania; 2003.
270. Morgan JR. Search for safety-lamp in mines. Annals of Science 1936; 1(3): 302–329.
271. Morris RA. Review of experiences on the use of inert gases in mine fires. Mining Science and Technology 1987; 6(1):37–69.
272. Morris R. Review of experiences on the use of inert gases in mine fires. Mining Science & Technology 1987; 6(1):37–69.
273. Morris WP. Development and administration of a mining company safety program. Bell Laboratories Record 1972; 52(8):40–44.
274. Mosinets VN. Ensuring radiation safety for uranium mine liquidation, mothballing, and reprofiling. Soviet Atomic Energy 1991; 70(5):355–360.
275. Mousa-Bahia AA, Coles GSV, Gibson MJ, Willett MJ. Characterization of semiconductor gas sensors 11: a critical comparison with respect to products of combustion. Sensors and Actuators 1994; 19(1–3):668–674.
276. Mousa-Bahia AA, Coles GSV. The characterization of semiconductor gas sensors 2: A critical comparison with respect to product of combustion. Proceedings of the Eurosensors VII Conference and Exhibition 1994:668–674.

277. Mudder T, Botz M. Cyanide and society: a critical review. European Journal of Mineral Processing and Environmental Protection 2004; 4(1):62–74.
278. Murphy JN. Coal mine health and safety research in the USA: The achievements of the US Bureau of Mines. Coal International 1994; 242(6):219–226.
279. Murphy JN. Instrumentation and automation to improve mine safety. Computers & Industrial Engineering 1976; 1:17–20.
280. Murtagh LT. Safety at faraday uranium mines, Canadian Mining Journal 1964; 85(9):66–70.
281. Naevestad TO. Safety understandings among crane operators and process operators on a Norwegian offshore platform. Safety Science 2008; 46(3): 520–534.
282. Newhouse TV, editor. Coal mine safety. Hauppage, New York: Nova Science Publishers; 2009.
283. Niu X, Huang X, Zhao Z, Zhang Y, Huang C, Cui L. The design and evaluation of a wireless sensor network for mine safety monitoring. Proceedings of the IEEE Global Telecommunications Conference 2007:1291–1295.
284. Nundlall AR. Case study: Lessons learnt from recent flammable gas explosions in South African hard rock mines. Journal of the Mine Ventilation Society of South Africa 2006; 59(2):64–69.
285. Nussey C, Worthington B, Pinkard D. The use of systematic reliability assessment techniques for the evaluation and design of mining electronic systems. Journal of Occupational Accidents1985; 7(1):1–16.
286. O'Dea A, Flin R. Site managers and safety leadership in the offshore oil and gas industry. Safety Science 2001; 37:39–57.
287. Ogden SO. Economics, safety, and operating advantages of diesels in underground coal mines. Advances in Electronics and Electron Physics, Underground Services and Operations 1978; 61(8):19–24.
288. Ostvik J, Rundmo T, Sjoberg L. Associations between safety climate and emotional reactions to platform movements onboard an offshore installation. Safety Science 1997; 26(3):155–168.
289. Pabst M. Changes aimed at workplace safety in Thompson mine. Canadian Mining Journal 1980; 101(11):21–29.
290. Paik JK, Seo JK. A method for progressive structural crashworthiness analysis under collisions and grounding. Thin-Walled Structures 2007; 45(1):15–23.
291. Palei SK, Das SK. Sensitivity analysis of support safety factor for predicting the effects of contributing parameters on roof falls in underground coal mines. International Journal of Coal Geology 2008; 75(4):241–247.
292. Pappas DM, Mark C. Profile of groundfall incidents in underground coal mines. Mining Engineering 2003; 55(9):65–71.
293. Paques JJ, Durka JL, Bourbonniere R. Practical use of IEC 61508 and EN 954 for the safety evaluation of an automatic mining truck. Reliability Engineering and System Safety 1999; 66(2):127–133.
294. Parker EW. Direct negotiation and open shop cited as safeguards of mine worker's freedom Coal Age 1920; 18(22):1086–1089.

295. Paton WB. "Safety" at Flin Flon. Canadian Mining Journal 1930; 51(41): 969–971.
296. Paton WB. Accident prevention. Canadian Mining Journal 1931; 52(4–25): 86–88.
297. Paton WB. Mine safety – Yesterday, today and tomorrow. Canadian Mining Journal 1938; 59(12):672–676.
298. Paul PS, Maiti J. The role of behavioral factors on safety management in underground mines. Safety Science 2007; 45(4):449–471.
299. Paul PS, Maiti J. The synergic role of sociotechnical and personal characteristics on work injuries in mines. Ergonomics 2008; 31(5):737–767.
300. Pearson TP. Mine safety and health administration and how it affects the synthetic fuels industry. IEEE Transactions on Industry Applications 1983: 19(6):1024–1030.
301. Pearson TP. The Mine Safety and Health Administration and how it affects the synthetic fuels industry. IEEE Transactions on Industry Applications 1983; 19(6):1024–1030.
302. Peters RH, Clingan MR. Reducing coal miner absenteeism. Mining Science and Technology 1989; 9(2):227–236.
303. Petrov NN, Barakhtenko EV. Ways of improving safety and reliability of ventilation in mines. Journal of Mining Science 2006; 60(2):171–177.
304. Platt SH. Electronic alarm systems make Selby a safer mine. Electrical Review 1983; 10(10):35–38.
305. Pontt J, Rodrigues J, Rebolledo J, San Martin J, Cid, E, Figgueroa G. High-power LCI grinding mill drive under faulty conditions. Proceedings of the IEEE 14th Industrial Applications Society Annual Conference 2005; 1:670–673.
306. Pugh JW, Johnson KS. Industrial-minerals health-and-safety inspections by the Oklahoma Department of Mines. Proceedings of the Conference on the Industrial-Minerals Development 1993:55–56.
307. Radomsky MC, Saperstein LW, Passmore DL, Bennett JD. Improving mandated training through on-site safety services. Mining Engineering, Littleton 1988; (6):444–446.
308. Ramani RV, Mutmansky JM. Mine health and safety at the turn of the millennium. Mining Engineering 1999; 51(9):25–30.
309. Ramani RV. Mining disasters caused and controlled by mankind; the case for coal mining and other minerals. Natural Resources Forum 1995; 19(3): 233–242.
310. Ramani RV, Mutmansky JM. Mine health and safety at the turn of the millennium. Mining Engineering 1999; 51(9):25–30.
311. Ramsay HT, Senneck CR. Anti-slip studs for safety footwear. Applied Ergonomics 1972; 3(4):219–223.
312. Rector III JW, Hardage JW. Radiation pattern and seismic waves generated by a working roller-crone drill bit. Geophysics 1992; 57:1319–1333.
313. Report of the Royal Commission on the Health and Safety of Workers in Mines. Ontario Ministry of the Attorney General, Toronto, Canada; 1976.

314. Reynolds RL. History of coal mine electrical fatalities since 1970. IEEE Transactions on Industry Applications 1985; 21(6):1538–1544.
315. Rice GS. Safety in coal mining: A handbook. Washington D.C.: Government Printing Office; 1928.
316. Robinson GH. Accidents and sociotechnical systems: principles for design. Accident Analysis & Prevention 1982; 14(2):121–130.
317. Ross E. Public address system in safety work at Perron gold mines. Canadian Mining Journal 1943; 64(8):512–513.
318. Ruff TM, Hession-Kunz D. Application of radio-frequency identification systems to collision avoidance in metal/nonmetal mines. IEEE Transactions on Industry Applications 2001; 37(1):112–116.
319. Ruff TM, Holden TP. Preventing collisions involving surface mining equipment: A GPS-based approach. Journal of Safety Research 2003; 34(2):175–181.
320. Rundmo T. Risk perception and safety on offshore petroleum platforms. Part II. Perceived risk, job stress and accident, Safety Science 1992; 15(1):53–68.
321. Rupert GB, Clark GB. Criteria for the proximity of surface blasting to underground coal mines. Proceedings of the Rock Mechanic Symposium 1977: 31–310.
322. Rusinski E, Kopcyzynski A, Czmochowski J. Tests of thin-wall beams joined by spot welding. Journal of Materials Processing Technology 2004; 157–158:405–409.
323. Sacks HK, Cawley JC, Homce G, Yenchek M. Feasibility study to reduce injuries and fatalities caused by contact of cranes, drill rigs, and haul trucks with high tension lines. Proceedings of the IEEE Industry Applications Society Annual Meeting. 1999; 1:240–246.
324. Safety and Health in Coal Mines. Report International Labor Office. Geneva, Switzerland; 1976.
325. Safety and Health in Opencast Mines. Report. International Labor Office. Geneva, Switzerland; 1991.
326. Safety in the Use and Maintenance of Large Mobile Surface Mining Equipment. Report. Bureau of Mines. United States Dept. of the Interior. Washington D.C.; 1983.
327. Sammarco JJ, Kohler JL, Novak T, Morley LA. Safety issues and the use of software-controlled equipment in the mining industry. Proceedings of the 32nd IEEE Industry Applications Annual Conference 1997:2084–2090.
328. Sammarco JJ. Programmable electronic and hardwired emergency shutdown systems: A quantified safety analysis. Proceeding of the 40th IEEE Industry Applications Society Annual Meeting 2005:210–217.
329. Sammarco JJ. Addressing the safety of programmable electronic mining systems: lessons learned. Proceedings of the 37th IEEE Industry Applications Society Conference Annual Meeting. 2002:692–698.
330. Sapko MJ, Cashdollar KL, Green GM. Coal dust particle size survey of US mines. Journal of Loss Prevention in the Process Industries 2007; 20(4-6):616–620.

331. Sari M, Duzgun HSB, Karpuz C, Selcuk AS. Accident analysis of two Turkish underground coal mines. Safety Science 2004; 42(8):675–690.
332. Sattel G, Sander BK, Amberg F, Kashiwa T. Predicting ahead of the face. Tunnels and Tunneling International 1996; 28(4):24–30.
333. Schiffbauer WH. Flexible control networks in mining machines to improve worker safety. IEEE Control Systems Magazine 1992; 12(3):87–94.
334. Schreiber W, Kielblock J. Self-contained self-rescuer legislation within the context of the Mine Health and Safety Act of South Africa: A critical analysis. Journal of the Mine Ventilation Society of South Africa 2004; 57(4): 119–123.
335. Schweisheimer W. Mine safety legislation through centuries. Canadian Mining Journal 1948; 69(6):92–93.
336. Seal AB, Bise CJ. Case study using task-based, noise-exposure assessment methods to evaluate miner noise hazard. Mining Engineering 2002; 54(11): 44–48.
337. Selby RW, Basili VR. Analyzing error-prone system structure. IEEE Transactions on Software Engineering 1991; 17(2):141–152.
338. Senneck CR. Over-3-day absences and safety. Applied Ergonomics 1975; 6(3):147–153.
339. Shaughnessy JM, Herrmann RP. Concentric riser will reduce mud weight margins, improve gas-handling safety. Oil and Gas Journal 1998; 96(44):5.
340. Shetty NK, Guedes Soares C, Thoft-Christensen P, Jensen FM. Fire safety assessment and optimal design of passive fire protection for offshore structures. Reliability Engineering & System Safety 1998; 61(1–2):139–149.
341. Simcox S, Wilkies NS, Jones IP. Computer simulation of the flows of hot gases from fires at King's Cross Underground Station. Fire Safety Journal 1992; 18(1): 49–82.
342. Singh AK, Singh RVK, Singh MP, Chandra H, Shukla NK. Mine fire gas indices and their application to Indian underground coal mine fires. International Journal of Coal Geology 2007; 69(3):192–204.
343. Singh PK. Blast vibration damage to underground coal mines from adjacent open-pit blasting. International Journal of Rock Mechanics and Mining Sciences 2002; 39(8):959–973.
344. Singh PK, Roy SJ, Sinha A. A new blast damage index for the safety of underground coal mine openings. Transactions of the Institution of Mining and Metallurgy 2003; 112(2):97–104.
345. Sjoberg L. Perceived risk and tampering with nature. Journal of Risk Research 2000; 3:353–367.
346. Smith E, Aclander M. Quarry operator fined in fatal fall case. Pit and Quarry 2003; 96(3):36.
347. Smith E, Aclander M, Commission further defines when an accident must be reported. Pit and Quarry 2003; 96(4):45.
348. Sneddon A, Mearns K, Flin R, Situation awareness and safety in offshore drill crews. Cognition, Technology & Work 2006; 8(4):255–267.
349. Snyder K. Health and safety. Mining Engineering 1984; 36(5):467–468.

350. Song HW, Lu SM. Repair of a deep-mine permanent access tunnel using bolt, mesh and shotcrete. Tunneling and Underground Space Technology 2001; 16(3):235–240.
351. Spokes EM. New look at underground coal mine safety. Mining Engineering 1986; 38(4):266–270.
352. Spokes EM. New look at underground coal mine safety. Mining Engineering 1985; 38(4):266–270.
353. Stacey A, Sharp JV. Safety factor requirements for the offshore industry. Engineering Failure Analysis 2007; 14(3):442–458.
354. Stefopoulos EK, Damigos DG. Design of emergency ventilation system for an underground storage facility. Tunneling and Underground Space Technology 2007; 22(3):293–302.
355. Sterling R, Carmody J, Rockenstein WH. Case study of life safety standards for a large mined underground space facility in Minneapolis, Minnesota. Tunneling and Underground Space Technology1992; 7(2):119–125.
356. Stevenson DH. Barrick improves safety at Goldstrike mines. Mining Engineering Littleton 1997; 49(1):42–44.
357. Stewart B, Iverson S, Beus M. Safety considerations for transport of ore and waste in underground ore passes. Mining Engineering 1999; 51(3):53–60.
358. Stoppe TJ, Bobick TG, Plummer RW. Musculoskeleton injuries in underground mining. Proceedings of the American Conference of Governmental Industrial Hygienists 1986:71–76.
359. Stormont D. Offshore rig designed for safety. Oil and Gas Journal 1959; 57(9):147–148, 151.
360. Stracher GB, Taylor TP. Coal fires burning out of control around the world: thermodynamic recipe for environmental catastrophe. International Journal of Coal Geology 2004; 59(1–2):7–17.
361. Sun ZQ, Zhang YR, He T, Yang CG. Expectancy of working life of mine workers in human province. Public Health 1997; 111(2):81–83.
362. Sun Y, Qian J. Forecast the dangerous level of gas outburst based on GASA neural networks in coal mine. Proceedings of the 3rd International Conference on Natural Computation 2007; 16:641–645.
363. Suphi U, Sitki D. Evaluation of occupational safety and health in surface mines. Safety Science 2008; 46(6):1016–1024.
364. Surendra M, Dayse D. Cognitive modeling of underground miners response to accidents. Proceedings of the Annual Reliability and Maintainability Symposium 2006:51–55.
365. Sutherland M. Improving mine safety through perimeter blasting at Inco Limited's Thompson T-3 Mine. Canadian of Institution Mining Bulletin 1990; 83(940):45–48.
366. Sutherland TF. A visit to gold mines of South Africa. Canadian Mining Journal 1926; 37(15):401–402.
367. Sven H. When safety matters: Using active RFID down the mines. Sensors 2004; 21(2): 34–39.

368. Sydanheimo L, Keskilammi M, Kivikoski M. Reliable mobile computing to underground mine. Proceedings of the IEEE International Conference on Communications 2000:882–888.
369. Sykes EH. Mine safety appliances. Canadian Mining Journal 1927; 48(43): 856–861.
370. Szwedzicki T. Quality assurance in mine ground control management. International Journal of Rock Mechanics and Mining Sciences 2003; 40(4):565–572.
371. Tango G. Safety in offshore drilling. American Association of Petroleum Geologists Bulletin 1992; 76(3):412–413.
372. Taschereau RH. Inspection of mines. Canadian Mining Journal 1948; 69(10):136–137.
373. Tesarik DR, Seymour JB, Yanske TR. Post-failure behavior of two mine pillars confined with backfill. International Journal of Rock Mechanics and Mining Sciences 2003; 40(2):221–232.
374. Thielemann T, Kross BM, Liltlce R, Welte DH. Does coal mining induce methane emission through the lithosphere/atmosphere boundary in the Ruhur Basin, Germany? Journal of Geochemical Exploration 2001; 74:219–231.
375. Thorne R. Reducing rock falls during mine rescue. Mining Engineering 1996; 48(12):15–18.
376. Tien JC. Health and safety challenges for China's mining industry. Mining Engineering 2005; 57(4):15–23.
377. Torbjorn R, Lennart S. Employee risk perception related to offshore oil platform movements. Safety Science 1996; 24(3):211–227.
378. Torma-Krajewski J, Steiner L, Lewis P, Gust P, Johnson K. Implementation of an ergonomics process at a US surface coal mine. International Journal of Industrial Ergonomics 2007; 37(2):157–67.
379. Trotter D. Safety's newest concern Em Dash Mine lighting. Canadian Mining Journal 1977; 98(7):24–25, 27–31.
380. Tveit OJ. Safety issues on offshore process installations: An overview. Journal of Loss Prevention in the Process Industries 1994; 7(4):267–272.
381. Udd JE, Betournay MC. Health, safety, and mine environments research in the mining laboratories of CANMET, Natural Resources Canada. Canadian Institution of Mining Bulletin 1997; 90(1010):45–51.
382. Ugur Y. Geoelectric exploration and monitoring in rock salt for the safety assessment of underground waste disposal sites. Journal of Applied Geophysics 2000; 44(2–3):181–196.
383. Urban K. Safety in the design of offshore platforms: Integrated safety versus safety as an add-on characteristic. Safety Science 2007; 45(1–2):107–127.
384. Urosek JE, Schade H. State of the art monitoring and computer analysis of the results following a recent major mine fire. Proceedings of the 4th US Mine Vent Symposium 1988:66–71.
385. Vanquelin O, Megret O. Smoke extraction experiments in case of fire in tunnels. Fire Safety Journal 2002; 37(5):525–533.

386. Vaught C, Mallett L, Brnich MJ, Reinke D, Kowalski-Trakofler, KM, Cole HP, Vinnem JE, Aven T, Husebo T, Seljelid J, Tveit OJ. Major hazard risk indicators for monitoring of trends in the Norwegian offshore petroleum sector. Reliability Engineering & System Safety 2006; 91(7):778–791.
387. Vinnem JE. Evaluation of methodology for QRA in offshore operations. Reliability Engineering & System Safety 1998; 61(1–2):39–52.
388. Vipperman JS, Bauer ER, Babich DR. Survey of noise in coal preparation plants. Journal of the Acoustical Society of America 2007; 121(1):197–205.
389. Viscusi WK. Wealth effects and earning premiums for job hazards. Review of Economic Statistics 1978; 60:408–416.
390. Vladislav K, Dragan K, Groves W, Radomsky M. An analysis of equipment-related fatal accidents in U.S. mining operations: 1995–2005. Safety Science 2007; 45(8):864–874.
391. Wachowicz J. Analysis of underground fires in Polish hard coal mines. Journal of China University of Mining and Technology 2008; 18(3):332–336.
392. Wade J. Escape artist: Mining rescue effort. Australian Mining 2006; 98(9): 6–12.
393. Wade J. In the frame for safer heavy vehicles: Heavy vehicle accidents in mining. Australian Mining 2006; 98(9):40.
394. Wade J. Is it safe? Safety performance of individual mine sites Australian Mining2007; 99(4):22–23.
395. Wala AM. Controlling ventilation for safe escape from coal mine fires. Mining Engineering, Littleton 1996; 69(4):61–66.
396. Wala AM, Dziurzynski W, Tracz J, Wooton D. Validation study of the mine-fire simulation model. Proceedings of the 7th US Mine Ventilation Symposium 1995:199–206.
397. Wang EY, Liu XF, Zhao EL, Liu ZT. Study of electromagnetic characteristics of stress distribution and sudden changes in the mining of gob-surrounded coal face. Journal of China University of Mining and Technology 2008; 18(1):1–5.
398. Wang J, Yang JB, Sen P, Ruxton T. Safety based design and maintenance optimisation of large marine engineering systems. Applied Ocean Research 1996; 18(1):13–27.
399. Wanger JP, Fookson A, May M. Performance characteristics of semiconductor sensors under pyrolytic, flaming, and smoldering combustion conditions. Journal of Fire Flammables 1976:71–103.
400. Watson AP, O'Donnell FR. Direct coal liquefaction safety risks: Feedstock production and transportation. Energy 1985; 10(7):877–886.
401. Watzman B. Safety and health opportunities and challenges for a resurgent mining industry. Coal Age 2004; 109(9):45–48.
402. Weber BH, Cooke JS, Borland RM, Watt A, Allen GA, Perry R J, Laing R., What price health, safety? Canadian Mining Journal 1972; 93(2):32.
403. Wepp WR, Peterson MO. Maintenance and training pay safety dividends for Benson mines. Mining Engineering.1952; 4(8):768–769.
404. Werniuk J. Fighting fire with training. Canadian Mining Journal 2006; 127(3):21–26.

405. Williams M, Schofield D, Denby B. Development of an intelligent haulage truck simulator for improving the safety of operation in surface mines. Lecture Notes in Computer Science 1998; 1434:337–345.
406. Winn FJ, Biersner RJ, Morrissey S. Exposure probabilities to ergonomic hazards among miners. International Journal of Industrial Ergonomics 1996; 18(5–6):417–422.
407. Wolinski TR, Zagorowicz J, Wyrwas AK, Domanski AW. Optical fiber temperature safety system for coal mines. Journal of the International Society for Optical Engineering 2002; 4887:67–71.
408. Wright CL. The first national mine-safety demonstration. Scientific American. December: 1911.
409. Wu Q, Wang M, Wu X. Investigations of groundwater bursting into coal mine seam floors from fault zones. International Journal of Rock Mechanics 2004; 557–571.
410. Yanjing S, Jinsheng Q, Shiyin L, Jinling S. Intelligent prediction system of coal-gas outburst based on evolutionary neural nets. Proceedings of the IEEE International Conference on Neural Networks 2006:2650–2655.
411. Yenchek MR. Bureau of mines shock-prevention research program – An overview. IEEE Transactions on Industry Applications 1986; 22(5):941–945.
412. Yu TR, Vongpaisal S. New blast damage index criteria for underground blasting. Canadian Institution of Mining Bulletin 1996; March:139–145.
413. Zeng W. Exploration for human factors in the design of coal-mine safety and rescue devices. Proceedings of the 7th International Conference on Computer-Aided Industrial Design and Conceptual Design 2006:412–416.
414. Zhang RX, Yu DF, Li XW, Yao XG, Liu Y. Surface mine system simulation and safety risk management. Journal of China University of Mining and Technology 2006; 16(4):413–415.
415. Zhijian X, Qiulin T. A mini multi-gas detection system based on infrared principle. International Journal of Infrared and Millimeter Waves 2006; 27(12): 1639–49.
416. Zhou F, Ren W, Wang D, Song T, Li X, Zhang Y. Application of three phase foam to fight an extraordinary serious coal mine fire. International Journal of Coal Geology 2006; 67(1–2):95–100.
417. Zhou FB, Ren WX, Wang DM. Application of three-phase foam to fight an extraordinarily serious coal mine fore. International Journal of Coal Geology 2006; 67(1–2):95–100.
418. Zhou FB, Wang DM, Zhang YJ, Zhang YL, Li X. Practice of Fighting Fire and Suppressing Explosion for a Super-Large and Highly Gassy Mine. Journal of China University of Mining and Technology 2007:459–463.
419. Zohar D. Safety climate in industrial organizations: theoretical and applied implications. Journal of Applied Psychology 1980; 65(1):96–102.
420. Zucker E. Safety and economy in mine electrification. Canadian Mining Journal 1961; 82(6):74–77.

Author Biography

Dr. B.S. Dhillon is a professor of Engineering Management in the Department of Mechanical Engineering at the University of Ottawa. He has served as a Chairman/Director of the Mechanical Engineering Department/Engineering Management Program for over 10 years at the same institution. He has published over 348 articles (i.e., 203 journal + 145 conference proceedings) on reliability, safety, engineering management, etc. He is or has been on the editorial boards of 10 international scientific journals. In addition, Dr. Dhillon has written 36 books on various aspects of reliability, design, safety, quality, and engineering management published by Wiley (1981), Van Nostrand (1982), Butterworth (1983), Marcel Dekker (1984), Pergamon (1986), etc. His books are being used in over 100 countries and many of them are translated into languages such as German, Russian, and Chinese. He has served as General Chairman of two international conferences on reliability and quality control held in Los Angeles and Paris in 1987.

Prof. Dhillon has served as a consultant to various organizations and bodies and has many years of experience in the industrial sector. At the University of Ottawa he has been teaching reliability, quality, engineering management, design, and related areas for over 30 years and he has also lectured in over 50 countries, including keynote addresses at various international scientific conferences held in North America, Europe, Asia, and Africa. In March 2004, Dr. Dhillon was a distinguished speaker at the Conf./Workshop on Surgical Errors sponsored by the White House Health and Safety Committee and the Pentagon and held on Capitol Hill.

Professor Dhillon attended the University of Wales where he received a BS in electrical and electronic engineering and an MS in mechanical engineering. He received a Ph.D. in industrial engineering from the University of Windsor.

Index

Breinigsville, PA USA
11 November 2010
249127BV00003B/2/P